Understanding
Dark Matter

ISBN: 1477539344
ISBN 13: 9781477539347
Library of Congress Control Number: 2012909622
CreateSpace North Charleston, South Carolina

Table of Contents

Understanding "Dark" "Matter"

1. Introduction

Now, at the beginning of the 21st century, problems have arisen in applying the precepts of the currently dominant theory of gravitation, General Relativity, to large-scale structures in the universe. Specifically, General Relativity is ineffective in at least two respects:

a. First, it fails to describe the motion of objects at virtually any distance from the center of galactic or larger systems unless one is willing to assume each galaxy or larger structure that has been analyzed contains -- and, indeed, is primarily composed of -- "objects" that have yet to be observed; and

b. Second, it fails to properly describe the behavior of large-scale purportedly expanding structures absent the assumption that the universe contains a form of energy that also has so far evaded detection.

The former problem has led to an extensive search for "objects" -- popularly identified as "Dark" "Matter" -- and the latter has led to an extensive search for a form of energy -- popularly identified as "Dark" "Energy" -- that together will preserve the dominant status of Einstein's theory of General Relativity as the approach to understanding gravitation.

The lengths to which physicists are willing to go to posit unobserved phenomena to preserve General Relativity as the theoretical framework for understanding gravitating systems is reminiscent of the lengths to which physicists were willing to go, in the 1800's and early 1900's, to preserve the notion of a lumiferous ether as a medium for the transmission of electro-magnetic phenomena and specifically as a medium that could be analyzed to develop an absolute frame of reference. The author believes it is time to again do what was done to move beyond problems with the electromagnetic ether -- to return to the conceptual roots of existing theories of mechanics and gravitation, both the Newtonian approaches and the approaches dictated by Special Relativity and General Relativity, to see whether there might be flaws which allow the replacement of the existing theory of gravitation with one that does not require the invention of unobservables. This work has undertaken such an analysis and purports to have found flaws that, when corrected, avoid the need to find what has thus far proven so elusive. Before describing these flaws, however, this work will briefly undertake to describe both the "Dark" "Matter" and "Dark" "Energy" problems.

2.   The Need for "Dark" "Matter"

As the 20th century advanced, the tools for astronomical observation developed rapidly. With these improved tools, astronomers implemented programs to infer the speed with which objects throughout a galactic structure move as compared to the motion of the galactic structure as a whole -- in other words, as compared to the motion of the center-of-mass of the structure[1]. Measurements of velocity distributions within a wide variety of galaxies have been made, and with few, if any, exceptions, these measurements demonstrate a trend that has confounded and still confounds expectations.

The expectations of astronomers, based on their analysis of the closest and most familiar gravitating system -- our Solar System -- were that objects distant from galactic cores should be rotating at a leisurely pace with respect to the relevant galactic centers. In our solar system, the orbital velocity of planets decreases inversely as the square root of r, the mean distance of the planet from the sun[2]. This property has

---

1   Note the role of the "center-of-mass" of a galactic system in the formulation of the "Dark" "Matter" problem. Foreshadowing what is to come, there is an inherent contradiction in applying a theory -- General Relativity -- that denies the existence of a unique frame of reference while, at the same time, posing a problem in the analysis of galactic and other large-scale structures that implicitly assumes that a unique, center-of-mass frame exists.

2   See Rubin, Vera C., *Dark Matter in Spiral Galaxies*, Scientific American, June, 1983, *reprinted in*, Hodge, Paul W., The Universe of Galaxies,

been known for centuries and is one of the foundations for Kepler's Laws. These laws, in turn, were a substantial part of the experimental evidence that led Isaac Newton to develop his theories of mechanics and of gravitation given the observed fact that more than 99 percent of the mass of the solar system can be found in the sun.

The relationship between orbital velocity and orbital radius is developed as follows. First, using Newton's mechanics, the centripetal acceleration of an object orbiting the sun is given to a reasonable approximation by the formula:

$$a_{centripetal} = \frac{(v_{planet})^2}{r_{planet's\ orbit}}$$

In the Solar System, the force that produces this acceleration is the force of gravity which, under Newton's classical approach, is given by the following formula:

$$F_{centripetal} = \frac{G(m_{planet})(M_{Sun})}{(r_{planet's\ orbit})^2}$$

Using Newton's Second Law, F = ma, it follows that:

$$\frac{(m_{planet})(v_{planet})^2}{r_{planet's\ orbit}} = \frac{G(m_{planet})(M_{Sun})}{(r_{planet's\ orbit})^2}$$

Since both sides of the equation can be divided by the mass of the particular planet involved, it becomes clear that the ratio of planetary velocity to planetary orbital radius is determined exclusively by the mass of the sun and does not depend at all on the mass of the planet. After we have eliminated the planet's mass in this manner, we can then multiply both sides of the equation by the radius of the orbit

(W.H. Freeman and Company, New York 1984). The statement in the text above is based on a table found at page 34 of the reprint.

of the planet involved to get a relationship between the square of the orbital velocity and orbital radius. After we take the positive square root of both sides of the resulting equation, we are left with the expected relationship between the planet's orbital velocity and the square root of its orbital radius.

Accordingly, in any system in which Newton's laws of motion and gravitation are reasonable approximations, we would expect the velocities of objects in orbit around a comparatively large, comparatively compact mass to vary inversely with the square root of their distance from that central mass, just as is true with the Solar System. Since astronomers initially assumed that there was a central concentration of mass in galaxies -- after all, most of the light from known galaxies is generated from galactic centers[3] -- it was assumed that the orbital velocities of outlying objects would decline in relation to the square root of their distance from that central concentration. In reality, the radial velocities of objects that are part of a galaxy appear initially to increase substantially with increasing distance from the galactic center, reach a peak or plateau that is not very far from the galactic center and then fall off only very slightly, if at all, with increasing distance. From this, astronomers have inferred that the total density of mass within a galaxy falls off very slowly as one moves away from the galactic center. This has set astronomers on a search for the non-luminous or barely luminous "objects" that account for this mass distribution[4]. These non-luminous and barely luminous "objects" are believed to account for the "Dark" "Matter" which scientists are currently pursuing.

3   See Rubin, Vera C., *Dark Matter in Spiral Galaxies*, Scientific American, June, 1983, *reprinted in*, Hodge, Paul W., The Universe of Galaxies, (W.H. Freeman and Company, New York 1984). The statement in the text above is based on language at the bottom of page 38 of the reprint.

4   See Rubin, Vera C., *Dark Matter in Spiral Galaxies*, Scientific American, June, 1983, *reprinted in*, Hodge, Paul W., The Universe of Galaxies, (W.H. Freeman and Company, New York 1984). The statement in the text above is based on language at the bottom of the middle column on page 31 of the reprint.

3.   The Need for "Dark" "Energy"

S till more recently, the tools for astronomical observation have been focused on large-scale purportedly expanding structures. These observations have led astronomers to infer that some unseen form of energy must exist to counteract the force of gravity within these expanding systems. According to the CERN Courier (available over the internet at some time during the preparation of this work):

> Dark energy was first discovered in 1998 by two groups using supernovae as markers of cosmological distance as a function of time -- the Supernova Cosmology Project led by Saul Perlmutter at Lawrence Berkeley National Laboratory and the High-z Supernova Search Team led by Brian Schmidt at Australian National University. Measurements indicated that distant supernovae were dimmer than expected from the cosmological inverse square law in a universe dominated by matter (S. Perlmutter, et al. 1999, A. Riess, et al. 1998). That is, they appeared to be further away than expected from the expansion rate of the universe if gravitation due to the matter contents were the main force. Some form of dark energy was required at the 99% confidence level, and in amounts sufficient to counteract, on cosmic scales, the gravitational attraction from the clustered matter.

> Since then, deeper and more precise supernova measurements and further lines of evidence confirm this conclusion (J Tonry et al. 2003, R Knop et al. 2003, D Spergel et al. 2003). Detailed

measurements of the cosmic microwave background power spectrum, by the Wilkinson Microwave Anisotropy Probe Satellite and by ground based experiments, imply the presence of dark energy too. They also show that the spatial geometry of the universe is consistent with the flatness prediction of inflation. But observations of galaxy clusters tell us that the matter contribution to the total energy density can amount to only 20-30% of the needed critical density. Any two of the three lines of evidence imply that the dark energy composes roughly three-quarters of he energy density of the universe, while the third method provides a cross check. Such an amount of dark energy acts to accelerate the cosmic expansion.

Those interested in a more detailed explanation as to why there is a need for "Dark" "Energy" are referred to the literature. The author's only comment on this issue at this point mirrors his views expressed in the introduction, above. The creation of yet another unobservable "dark" component of the universe suggests a comparison between the attitudes of physicists of the late 19th-century and the attitudes of physicists of the late 20th-century. The earlier group developed a myriad of theories to explain their failure to detect absolute motion by analysis of electrodynamic systems. The latter has developed a suite of unobserved universal components to explain why their existing gravitational theory remains valid despite its repeated failure. As before, what is needed goes far beyond refinements of existing concepts. Instead, absolutely everything needs to be re-examined.

4. The Author's Proposal for Dark Matter -- Gravitational Potential Energy

To the author, the "answer" to the problem of "Dark" "Matter" seems surprisingly intuitive. The separation of any pair of objects in the universe requires a quantity of energy that is based both on the masses of the discrete objects *and also* the amount of the separation. This energy quantity has mass that cannot be associated with or integrated into either object. "Dark" "Matter" within a galaxy, then, must include the sum total of the interaction energies of all of the pairs of objects that make up that galaxy, and, indeed, must consider all of the interaction potential energy implied by the arrangement of objects in that galaxy. The further out from a galactic center an object is located, the more potential energy is operative between that object and the galactic center and the greater the gravitational influence of this potential energy -- as opposed to interior celestial bodies -- on the behavior of the relevant object.

The theoretical basis for the author's intuition is certainly not revolutionary. One of the clear teachings of Special Relativity is that energy, in any form, must have a mass equivalence. Indeed, Albert Einstein's writings make clear his belief that gravitational potential energy must, itself, exert a gravitational influence.[5] Thus, a large-scale gravitating

---

5   See, e.g., A. Einstein, The Foundation of the General Theory of Relativity, §16, *reprinted in* The Principle of Relativity, Dover Publications, Inc. (1952), toward the bottom of the section (at the top of page 149), where it states "It must be admitted that this introduction of the energy-tensor

structure *inevitably must* have a mass greater than the mass of the discrete "objects" that make it up and, at the fringe of such a large-scale structure, the dominant role in determining gravitational force is likely to be the pull of system potential energy rather than the pull of the constituent discrete objects. Objects at the edge of a large-scale system are but the "tips" of icebergs. Their mere presence at a point distant from the system center-of-mass requires the presence of immense amounts of gravitating potential energy between them and that center-of-mass and it is the gravitational influence associated with this potential energy and the potential energy implicit in the location of other objects within the system that largely determines the behavior of objects at the fringe of the system[6].

---

of matter is not justified by the relativity postulate alone. For this reason we have here deduced it from the requirement that the energy of the gravitational field shall act gravitatively in the same way as any other kind of energy." See also, L. Brown, A Pais and B. Pippard, Twentieth Century Physics, Institute of Physics Publishing and American Institute of Physics Press (Bristol, Philadelphia and New York, 1995) at Section 4.3.2 on page 291:

> Einstein developed a scalar field theory of gravitation based on the equivalence principle, first treating the case of a static field. Having previously (1911) derived gravitational light deflection by taking the gravitational potential as an effective index of refraction, which causes the speed of light to vary from point to point, he now (1912) took this varying speed of light $c(r)$ as the scalar field representing gravitation. He argued that since gravitational energy has inertial mass ($m = E/c^2$) and inertial mass equals gravitational mass, gravitational energy must be one source of the gravitational field. Since the gravitational field acts as one of its own sources, $c(r)$ must obey a non-linear equation. This feature of non-linear self-interaction proved to be an abiding feature in the later development of his gravitational theory.

6   A reasonable analogy is to our current number system. The numbers 951.159 and 159.951 obviously have the same digits but we recognize the former as by far greater because of the concept of "place value." By convention, the further to the left of the decimal point a digit is, the greater its contribution to a number's value. Each place to the left of the decimal

implies multiplication of the digit found there by 10 raised to the power equal to the number of places to the left the digit is found. In gravitating systems, the decimal point would be replaced by the point that is the center-of-mass of the system. We will discuss below a method to assign a multiplier to each place further away from this center but it should be clear with only modest contemplation that, the further removed from the center-of-mass an object is, the greater the object's energy of position within the system and the greater quantity of potential energy is present in the system. Two systems with identical components but with greater separations between the system center-of-mass and the component objects or with the more massive components substituted for the less massive components at the greater distances (say in the following order: mass of 9, mass of 5, mass of 1, center-of-mass, mass of 1, mass of 5 and mass of 9 as compared with the following order: mass of 1, mass of 5, mass of 9, center-of-mass, mass of 9, mass of 5, mass of 1) simply cannot have the same aggregate gravitational influence so long as energy has a mass equivalence. The mass equivalence of all forms of energy, of course, is one of the most well tested of all of the mechanical concepts of Special Relativity. To further expand on this idea, consider the series of characters "|||" -- a contemporary mind likely would see these characters as the number "one hundred eleven" given the place value concept that we currently employ. To an ancient Roman, however, those characters are the number "three". The Roman numbering system treats the figure "|" as one item regardless of its location with respect to other numbers (save of, course, when to the left of a numeral of greater value when its position gives it a negative worth). The current approach to the construction of gravitating systems similarly gives no consideration to the question of position in determining "field" strength. To be sure, no amount of arguing would convince an ancient Roman that his or her identification of the value of the number was "wrong". Number systems are simply conventions rather than experimental facts, so that the use of a place value convention is a matter of taste. On the other hand, if one puts a weight marked with the characters "|||" on a scale and the scale reads "one hundred eleven pounds", one could say with certainty that a place value approach was incorporated into these marks and that the Roman numeral system was not used in the marking of this weight. The experiments that document the "dark" "matter" problem are analogous to this simple weighing and leave no doubt that an approach to gravitation based on the simple addition of fields of discrete objects without consideration of their position in an overall system cannot reflect

The real mystery to "Dark" "Matter", then, is not what it is -- it is gravitational potential energy -- but why so many individuals with great sophistication in physics, including Albert Einstein himself, are so very surprised to see its influence[7]. To solve this mystery, one need only look

---

reality. There must be a place value element to a proper theory of gravitation yet there is no place value element in the current theory. For this reason (and ultimately for a variety of others as discussed at length below) the current theory cannot be correct.

7 The only physicist the author is aware of whose writings seem to suggest an appreciation of the problem is Leon Brillouin. The second chapter of his book, Relativity Reexamined begins with a Section on "Relativity and Potential Energy." The discussion there recites the formula for the interrelationship of energy and mass that is popularly associated with Albert Einstein, $E = mc^2$, but Mr. Brillouin continues by noting that there is general confusion regarding the mass equivalence of potential energy. Thus, as observed by Mr. Brillouin, the objects and energy contained in a closed system, be that energy chemical, mechanical, kinetic or potential, have a mass equivalence, M_0. When the system is set in motion with velocity, v, the original mass equivalence -- including the mass equivalence of the internal potential energy -- is enhanced with the enhancement calculated using the formula:

$$M = \frac{M_0}{(1 - v^2/c^2)^{1/2}}$$

Mr. Brillouin notes, however, that the system may be part of a larger system, and may, by virtue of its position in this larger system, have an external potential energy not originally considered in the calculation of its now subsystem rest mass. When some of this formerly external potential energy becomes internalized as a result of increases to the velocity of the subsystem, it must now have a positive mass equivalence. Previously, it had none and, thus, the enhancement of the subsystem mass associated with its more rapid motion appears to come from nowhere. See L. Brillouin, Relativity Reexamined, Academic Press (1970), Chapter 2, at pages 13 and 14. The author, therefore, is simply expanding on the misgivings of Mr. Brillouin whose credentials as a serious scientist cannot really be doubted. Indeed, the author can not only confirm that Mr. Brillouin's misgivings were proper but can go further and show that

through standard introductory texts on mechanics and gravitation. The ideas in those texts, especially the concepts of mass, energy and local mechanics, belong to a by-gone era -- many go back almost to the time of Isaac Newton himself. Our current concepts of mass, energy and local mechanics, however, largely date from the time after Einstein's Special Relativity paper was published in 1905. As discussed in detail below, the author believes that the architects of our current understanding of gravitation retained many assumptions considered so well-established prior to 1905 as to be beyond question, not realizing that those assumptions are demonstrably at odds with the mechanics dictated by Special Relativity.

Among the problem assumptions to be discussed at this point[8], as developed by the author from a review of a wide variety of texts, are as follows:

- That one may use Newton's law of gravitation to determine the behavior of two discrete objects, and, in determining the potential energy of the resulting gravitating system, one may add an *"arbitrary"* constant.

- That a permissible (if not mandatory) choice of this *"arbitrary"* constant is *zero* for two objects, regardless of their mass, when these objects are *an infinite distance from each other*;[9] and

---

Mr. Brillouin's misgivings ultimately lead to the answer to the question of what is "Dark" "Matter." Mr. Brillouin likely would have found the answer himself if he had but recognized the significance of the erroneous setting of the zero-point of gravitational potential energy discussed in Section 5.c., below. Unfortunately, he did not. See L Brillouin, Relativity Reexamined, Academic Press (1970), at figure 6.1 on page 78.

8   Section 8 of this work discusses a variety of additional concepts.

9   See, R. Weidner and R. Sells, Elementary Classical Physics, Allyn and Bacon, Inc. (1973), starting under section 13-5 at page 246-247.

◆ That the potential energy in a gravitating system is a *nega-tive* quantity that is classically calculated by the formula[10]:

$$(\text{Potential Energy})_{Gravitational} \quad = \quad - \quad \frac{G(M_1)\,(M_2)}{(r_{1,2})}$$

---

10 This formula is found in Section 13.6 at page 404 of J. Jewett, Jr. and R. Serway, <u>Physics for Scientists and Engineers</u>, 6th Ed. (Thompson Brooks/Cole 2004). In the formula, G is the gravitational constant, M_1 and M_2 are the masses of the relevant objects and r is the distance between the centers of the two objects and thus twice the radius of either object if the objects are identical and are just touching each other.

5. Analysis of Problem Concepts Incorporated into the Current Understanding of Gravitation

a. Precepts Assumed as Valid

In order to test the alleged problem assumptions outlined above against the precepts of Special Relativity, one must first select those precepts of Special Relativity and related concepts of physical systems that are to be assumed valid and used in each test. The author has selected the following:

First, the now familiar equation:[11]

$$E = mc^2$$

although often in the analysis that follows we will apply this formula after it has been algebraically manipulation into the following form:

$$E/c^2 = m \quad \text{or} \quad m = E/c^2$$

Second, that the inertial mass of an object that is moving with respect to a particular frame of reference is enhanced beyond its rest

---

11 See, e.g., French, A.P., Special Relativity, W.W. Norton & Company, Inc. (1968), Equation 1-8 at page 17, which includes both variants, mass in terms of energy and energy in terms of mass, of the relevant equality.

mass, m_0, with the enhanced mass in this particular frame determined by use of the following formula:[12]

$$m(v) \quad = \quad \frac{m_0}{(1-v^2/c^2)^{1/2}}$$

Third, that gravity is a conservative force, so that we can construct an arrangement of matter by any means and believe that the potential energy in this gravitational system is a single measurable quantity independent of the manner in which the arrangement was constructed.[13]

---

12 See, French, A.P., <u>Special Relativity</u>, W.W. Norton & Company, Inc. (1968), Equation 1-18 at page 22; and W.G.V. Rosser, <u>An Introduction to the Theory of Relativity</u>, Butterworth & Co. (Publishers) LTD. (1964) at Section 5.2 beginning on page 175 and Sections 5.6 and 5.7 beginning on page 207.

13 See, D. Tilley, <u>University Physics for Science and Engineering</u>, Cummings Publishing Company, Inc., Menlo Park, California (1976), at section 14.2 and especially the portion starting at page 228 where the following discussion (with emphasis added by the author) appears:

> **Internal Gravitational Potential Energy of a System**
>
> A system comprising several particles has an internal gravitational potential energy associated with each pair of particles due to their gravitational interaction. Within the system, a pair of particles of masses m_1 and m_2 separated by a distance r_{12} has a gravitational potential energy:
>
> $$U_{12} \quad = \quad - \quad \frac{Gm_1m_2}{r_{12}}$$
>
> ___This internal gravitational potential energy belongs to the pair and is not assigned to one specific particle of the pair.___

Following this discussion is an extremely interesting example, Example 5, involving two identical spheres, each of mass m and diameter D initially separated by a distance r. Their gravitational interaction causes these spheres to accelerate toward one another such that, just before they collide, their velocities are each:

$$v \quad = \quad \{Gm\,[(1/D) \quad - \quad (1/r)]\}^{1/2}.$$

The author finds this example significant because it indicates that one must specify both the masses of the objects *and their separation* to determine the quantity of gravitational potential energy in a system. Once it is recognized that this potential energy has a mass equivalence, it should be clear that there are no two body gravitational problems. There are only three body problems -- two separated objects and the mass equivalence of the energy implicit in their separation. Any approach to gravitation that simply adds the rest energies of the two objects and does not take into account the energy implicit in their separation will inevitably fail when the energy of separation becomes an important factor. Indeed, upon very little reflection it should be obvious that the classical gravitational example from Mr. Tilley's text will inevitably result in the relativistic collision discussed in the second of our assumed precepts. The result of that collision, of course, is an object more massive than the sum of the objects at rest by the mass equivalence of the kinetic energy of these objects in their center-of-mass reference frame. This kinetic energy, in turn, was determined the moment we fixed the mass of each **and** the distance separating the masses.

See also, on the uniqueness of the quantity of gravitational potential energy in each system, D. Klepper and R. Kolinkow, Introduction to Mechanics, W.W. Norton & Company, Inc. (1968), starting under section 4.6 at page 162 and continuing at section 4.7 on page 168. See as well, E. Purcell, Electricity and Magnetism, (McGraw-Hill Book Company 1965) at section 2.8 beginning at page 51 with particular emphasis on the statement: "[t]he potential energy of a system of charges, which is the total work required to assemble the system, can be calculated from the electric field itself simply by assigning an amount of energy ($E^2/8\pi$) dv to every volume element dv and integrating over all space where there is electric field". This specific statement, of course, is applied in the context of assembly of a system of electrical charges but there is no reason at this point in our discussion that it should not also be true of an assembly of gravitating objects. See also, R. Weidner and R. Sells, Elementary Classical Physics, Allyn and Bacon, Inc. (1973), Section 8-4, pages 116-118 and Section 10-4, page 162. Indeed, if our third precept is not true of gravitating systems, then one of the most basic of

Fourth, that mass/energy is conserved.[14]

---

conservation laws -- the conservation of mass/energy -- essential to the development of the mechanics of Special Relativity cannot be defended. On the other hand, there is an inherent problem with the simultaneous assumptions that the potential energy in a gravitating system is a single quantity and that all observers would reach equal conclusions as to the amount and sources of this potential energy. This is a consequence of the variation of mass with velocity required by the mechanics of Special Relativity. To illustrate the problem, imagine two observers of unequal mass set in motion by a single event -- a bullet fired from a gun and the recoiling much more massive shooter. The velocities of these two "observers" must be different and thus their measurements (using the Lorentz transformation equations appropriate to their rest frames of reference) of the masses of all the discrete objects in the universe (and of the distances between those objects) could not agree. These observers, therefore, could not agree on the mechanism by which the purportedly unique gravitational "field" they experience came into existence. Imagine further that, instead of a shooter firing a gun, one has a system that consists of an electron accelerated in a modern particle accelerator and the elaborate acceleration mechanism. Here the accelerated particle moving at a relativistic velocity would attach fantastic values to all of the masses in its "universe" but the accelerating system would see the universe as little changed. Finally, imagine a system consisting of an observer turning on a flashlight. One of the two classes of interlinked "objects" -- the photons from the flashlight -- would see all of the masses in the universe as infinite.

14 Note that it should be possible, based on this concept, to measure the mass/potential energy stored in a gravitational "field" by assuming that an object at a specific distance -- at this point, we will not attempt to determine whether this distance is to be computed applying the concepts of Euclidian or of Riemanian geometry -- from a mass distribution arrived at its present location by being ejected outward from the center-of-mass of the distribution with just enough kinetic energy to reach this distance and by determining the positive mass equivalence associated with this kinetic energy (or no less than the mass equivalence of the fuel necessary to produce this kinetic energy) at the point of departure. Calculations of this sort will be central to the discussion that follows. The author sees no flaw in such an approach given the oft-stated assertion of relativists

Fifth, the gravitational mass and inertial mass of an "object" are identical.[15]

(and, of course, overwhelming experimental proof) that all energy has a mass equivalence. If physics is to preserve its most fundamental conservation laws -- initially conservation of mass and conservation of energy and now conservation of mass/energy -- then the amount of mass/energy in any gravitating system must be the sum of all "mass" plus the sum of all "energy" (converted to its mass equivalence). Indeed, as discussed below, the failure of physicists to convert all quantities of energy into a positive mass equivalence when attempting to solve gravitational problems appears the root of all of the "dark" mysteries in physics. Further, as intimated in footnote 1, above, the author doubts that physicists can legitimately question the use of a master, "center-of-mass" frame of reference for the inventory of all "mass" and all "energy" even for very large systems. Those who have framed the "dark" "matter" problem seem comfortable with the idea, even though the author's understanding of General Relativity would never allow the same level of comfort, and, of course, the author is not alone in sensing general confusion regarding how General Relativity operates. See L. Brown, A. Pais and B. Pippard, Twentieth Century Physics, Volume 1, Section 4.3.9 at page 301. As we will see, however, the problems do not lie with the idea of a "center-of-mass" reference frame for analyzing large systems but with those tenets of General Relativity that imply the impropriety of such a frame. In this regard, see Rindler, W., Essential Relativity, at Section 5.7 (Springer-Verlag 1977), where Mr. Rindler appears to establish: (1) that there is a unique center-of-momentum frame of reference for any system of objects; (2) that this center-of-momentum frame is also the center-of-mass frame: (3) that the mass in the center-of-mass or center-of-momentum frame corresponds to the "rest" mass of the system if its composite nature were not recognized; and (4) that observers on all objects in the system would agree on the "mass" and "velocity" of the center-of-momentum frame. Accordingly, if what Mr. Rindler has proven is, indeed, true (and the author thinks it is), then Mr. Rindler has disproven the central tenet of both General Relativity and Special Relativity -- that all frames of reference are equivalent. A further discussion of this point is later in this work.

15 This, of course, is a critical concept in the development of General Relativity. It has been retained here as true of what we typically consider "objects" in the weak gravitational "fields" that are the limit of human

b. The Problem with Adding an "Arbitrary" Constant

The first problem statement identified above is that one may use Newton's law of gravitation to determine the behavior of two discrete objects and, in determining the potential energy of the resulting gravitating system, one may add an "arbitrary" constant. When energy and mass were considered to be distinct and separate phenomena, this statement was absolutely true. It is basic calculus that, if one integrates the work done over a change in position by a force that is exclusively a function of two fixed masses and their relative position, one must add a constant to the result and must have more information to fix this constant. Similarly, if one differentiated a gravitational potential function one had constructed based on Newton's law of gravitation as applied to two discrete objects, one would expect to be able to ignore any constant amount of background energy.

On the other hand, if it is recognized that the constant chosen necessarily has a mass equivalence -- a fact that is an inescapable consequence of Special Relativity but that was unrecognized prior to 1905 -- then one must recognize that the addition of any non-zero constant is, in reality, the addition of a third gravitating mass to the problem that will prevent rigorous application of Newton's simple gravitational force law. Moreover, if one asserts that the amount of the constant is *arbitrary*, one encounters an insoluble problem because this third gravitating quantity can be chosen to be 1, or 2, or 3, or any arbitrary number of orders of magnitude more massive than any or all of the discrete masses that one is attempting to analyze. All that must be done is to multiply the mass of the largest of the objects or the sum of all masses that are being analyzed by 10 or 100 or 1000 (or more) and

---

experience. Note, however, that, for reasons discussed at length below, the concept of an "object" will need to be reconsidered as we examine the behavior of electro-magnetic fields in large systems.

then convert the resulting mass figure to a quantity of energy using the second form of the first of the equations referenced in 5.a., above.

Thus, consider an analysis of two neutrons originally just touching each other[16]. The mass of a single neutron is 1.674 927 16 x 10^{-27}Kg[17]

---

16 Available information, of course, suggests that a neutron is not a stable elementary particle so that the thought experiment described here -- like many other thought experiments to be sure -- could not be performed. Further, the radius of a neutron is not known with precision and, moreover, quantum mechanics suggests that it would not be possible to know that radius with confidence. The use of two neutrons, nevertheless, ties in with the further discussion below. For those that demand greater precision in their thought experiments, the author suggests an analysis of two hydrogen atoms, which are believed to be stable, and their gravitational "collapse" into a diatomic hydrogen molecule (or their elastic collision followed by their movement out to infinite separation). The numbers will change but the final concept will be the same -- that gravitation will result in the acceleration of the two discrete objects or systems toward each other until there is a collision; that there will be a positive amount of kinetic energy at the time of the collision; that this kinetic energy has a positive mass equivalence; and that this kinetic energy, when previously existing as potential energy and when converted back to potential energy as the two objects move away from each other against their mutual gravitational attraction will have associated with it an equally positive mass all as measured in the system's center-of-mass frame of reference.

Also, as an interesting aside (that the author believes will prove extremely important later) it is well to note that the instability of neutrons means that there are no neutral objects in the universe that are both stable and have a rest mass. A neutron can be at rest but is not stable and photons and perhaps neutrinos as well are arguably stable but cannot be at rest. The universe does not simply include electrically charged objects. It is made of them and only of them.

17 See J. Jewett, Jr. and R. Serway, Physics for Scientists and Engineers, 6th Ed., at the inside front cover (Thompson Brooks/Cole 2004)

and the radius of a neutron is approximately 1.0×10^{-15}m.[18] If we insert this information into the standard formula from Newton's theory of gravitation for determining the potential energy of two masses as a function of position[19], we get:

$$U \; = \; - \; \frac{[6.673 \times 10^{-11} \text{ N m}^2/\text{Kg}^2]\,[(1.674\,927\,16 \times 10^{-27}\text{Kg})]^2}{2 \times (1.0 \times 10^{-15})\text{m}} \; + \; C\text{(in Joules)}$$

Simplifying, we get

$$U \; = \; - \; 9.36 \times 10^{-50} \text{ N m} \; + \; C \text{ (in Joules)}$$

or

$$U \; = \; - \; 9.36 \times 10^{-50} \text{ Joules} \; + \; C \text{ (in Joules)}$$

Using the second form of the first of the formula referenced in 5.a., above, this equation can be converted to its mass equivalence as follows:

$$U \; = \; - \; \frac{9.36 \times 10^{-50} \text{ Joules}}{(2.997\,924\,58 \times 10^{8} \text{ m/sec})^2} \; + \; \frac{C\text{(in Joules)}}{(2.997\,924\,58 \times 10^{8} \text{ m/sec})^2}$$

---

18 See Wikipedia, the free encyclopedia.

19 This formula, of course, is derivative of the third of the problem concepts identified above. As noted there, the formula can be found in Section 13.6 on page 404 of J. Jewett, Jr. and R. Serway, Physics for Scientists and Engineers, 6th Ed. (Thompson Brooks/Cole 2004), and is as follows:

$$U \; = \; - \; \frac{Gm_1 m_2}{r}$$

In this formula, G is the gravitational constant, m_1 and m_2 are the masses of the relevant objects and r is the distance between the centers of the two objects which is twice the radius of either object here because the objects are the same.

$$U = - \frac{9.36 \times 10^{-50} \text{ Joules}}{8.987\ 551\ 787 \times 10^{16} \text{m}^2/\text{sec}^2} + \frac{C \text{(in Joules)}}{8.987\ 551\ 787 \times 10^{16} \text{m}^2/\text{sec}^2}$$

$$U = - 1.04 \times 10^{-66} \text{ Kilograms} + C_{Kg} \text{ (in Kilograms)}$$

Accordingly, if we pick our arbitrary energy constant at $8.987\ 551\ 787 \times 10^{16}$ Joules, we have really added 1 kilogram of potential energy to our system and our constant is 27 orders of magnitude greater than the mass of either of the objects we were attempting to analyze and 66 orders of magnitude greater than the potential energy of separation of those objects. Any determinations we may have made as to the behavior of the two objects would no longer be valid if the constant amount of energy we have arbitrarily added were indeed present anywhere in the "universe" we were considering[20]. Our two objects would behave in a manner dictated by this "arbitrary" energy constant and would largely ignore each other.

Ultimately, then, we cannot add an arbitrary amount or, indeed, any amount of energy to a two-object gravitating system and then assert that our purportedly rigorous analysis of that system is anything more than just a very local and very rough approximation. To rigorously analyze such a system, one must be meticulous in selecting and respecting its natural energy zero-point. That natural zero-point and its significance are discussed in the next section below.

---

20 The only way that the energy constant can avoid having an influence on the motion of the two objects is if the energy exerts a pull that has no origin. This kind of sourceless "field" is unknown in existing conceptions of gravitation.

c. The Problem with Defining the Zero-Point of Gravitational Potential Energy at Infinite Separation

The author's arguments challenging the first of the problem statements above are just quibbles when compared with concerns regarding the Newtonian selection of infinite separation as the zero-point for measuring gravitational potential energy. Once one sees the pernicious effects of this simple ***convention*** -- this idea is, after all, only an agreement among physicists and nothing more -- one knows with confidence that the existing system of gravitational analysis is deeply flawed and cannot be applied to large-scale structures in the universe. Logic leads inescapably to the selection of the point of closest approach as the natural potential energy zero-point in any gravitating system. Certainly it takes energy to separate objects in space and one can use the energy of objects returning to a gravitating body such as the earth to perform all manner of useful (or damaging) work[21]. Human society

----

21 Note that, in present thought, the zero-points of potential energy of electrical systems of like charge and gravitating systems are *the same* -- we ***define*** the potential energy of *both* such systems to be zero at infinite separation. The reason given for setting the zero-point of the potential energy of same-charge electrical systems at infinite separation is that "[t]his is consistent with the fact that positive work must be done by an external agent on the system to bring the two charges [of like sign] near one another (because charges of the same sign repel)." See J. Jewett, Jr. and R. Serway, Physics for Scientists and Engineers, 6th Ed., Section 25.3 with emphasis on the language on page 770 (Thompson Brooks/Cole 2004). This rational is obviously sound. The real reason behind the often inarticulately stated justifications offered for setting the gravitational zero-point at infinite separation is, to a great extent, that physicists have always done it that way. Flawed concepts apparently have a form of inertia that may be more powerful than the inertia of the largest of moving things. To confirm this fundamental inconsistency, compare the diagrams of potential energy of electrical systems of like charges and opposite charges at section 23.1, Figure 23-1 on page 401 from D. Tilley, University Physics for Science and Engineering, Cummings

Publishing Company, Inc., Menlo Park, California (1976) with the diagram of potential energy of gravitational systems from the same book at Section 14.2, Figure 14-5 on page 227. The diagram of the gravitational potential energy should be the same as the diagram of charges of opposite sign but actually the gravitational diagram is identical to the diagram for charges of like sign. It is clear, then, that the electrical and gravitational approaches are inconsistent and irreconcilable. This error, of course, was not originally an "error" at all but simply a choice between two apparently available alternatives. The continued survival of the pre-20th century approach is merely an anachronism -- consistency was unnecessary to physicists prior to the dawn of the last century when it was assumed that the work required to create something from diffuse electrically charged components or from a compact arrangement of gravitating objects had no gravitational influence. Now that it is clear that all work done against electrical repulsion or gravitational attraction creates potential energy and that this potential energy has mass and thus a gravitational influence, consistency is required and the only correct approach is to start with system subparts in their lowest energy state. For charges of like sign, this is at infinite separation. For masses (which can have only one sign) it is at the point of an infinitely compact point mass. This, to be sure, poses complementary problems. For electrical systems, the problem has always been to explain why the most basic electrical systems -- electrons and protons, for example -- stay together. For gravitating systems, the problem -- and perhaps the most significant of the reasons why our distant predecessors in the study of gravitating systems picked infinite separation as their zero-point -- has always been to explain why such systems don't collapse. Ultimately, these "problems" may each solve the other. The expansion of an electron to a greater radius as a result of the electrostatic repulsion of its constituent parts necessarily diminishes the mass-equivalence of the potential energy stored in the overall electric field -- because it eliminates the energy stored in the space in which the field has ceased to exist as a result of expansion. See E. Purcell, Electricity and Magnetism, (McGraw-Hill Book Company 1965) beginning at page 49. Conversely, the electrical potential energy implied by a reduction in the radius of an electron -- the energy cost of shrinking the electron and thus creation of an electric field in the space from which an electrical field was formerly lacking -- adds a new mass element to the universe in the form of the energy now stored in the newly created electric field. See again, E. Purcell, Electricity and Magnetism, (McGraw-Hill Book Company 1965) beginning at page 51 with particular emphasis on the statement: "[t]he

generates tremendous quantities of hydro-electric power by harnessing the energy of water lifted to great heights by solar energy as it returns to the earth's oceans and no observer would lightly stand beneath a piano falling to the earth from a third story window lest the substantial positive kinetic energy the piano obtains during the course of its fall do all manner of frightening "work" on the observer as it comes to a stop.

When the "zero-point-at-infinite-separation" convention was adopted, of course, the choice appeared arbitrary and the selection represented a simplification. This zero-point setting meant that the sum of the positive kinetic energy of a body and the negative potential energy associated with the position of this object in a gravitating system would always total zero and that the gravitational potentials of all

---

potential energy of a system of charges, which is the total work required to assemble the system, can be calculated from the electric field itself simply by assigning an amount of energy $(E^2/8\pi)dv$ to every volume element dv and integrating over all space where there is electric field". See also, D.G. Ivey and J.N. Patterson Hume, Physics (In Two Volumes), Volume 2 (The Ronald Press Company New York 1974) at pages 150-151. The energy required to shrink an electron into a smaller and smaller space tends to infinity. Accordingly, the requirement of energy conservation necessarily provides what is needed for electron cohesion – an electron cannot expand without the transfer of its field energy to some other portion of the universe. Correspondingly, there is no problem with gravitational collapse because mass is simply an attribute of the energy stored in electromagnetic fields and Gravitation is not a force that exists separate and apart from electricity and magnetism. Instead, Gravitation is merely evidence of the conserved nature of the energy and therefore mass stored in electromagnetic fields and the fundamental constraint on local transformations of the electrical-magnetic potential requiring that changes in the energy content of the field at a particular point in space and at a particular time be counterbalanced by changes at other places and at the same time (with, of course, all calculations done in one consistent frame of reference, but bearing in mind that the transformation of the field from one frame to another be done with the Lorentz transformation equations rather than the classical Galilean transformation equations). Obviously, the author will have more to say on this issue later.

objects, regardless of mass, would be identically zero at infinity[22]. The significance of the mass of the object being analyzed only determined the depth of the energy well into which other objects might fall. This convention also makes the zero-point of potential energy consistent with the point of zero force[23]. Further, this convention makes it unnecessary to know the entire inventory of the universe in determining the potential energy at any particular point. Moreover and perhaps most significantly, to physicists of the 17th, 18th, and 19th centuries, this zero-point avoided touchy fundamental questions -- when would an object falling under the influence of gravitation stop and why? We now have better answers to the questions of when and why, but the repulsive character of same charge electro-static interactions, the fact that all material objects consist of charged bodies in delicate balance and the inability of gravitation to overcome this balance -- at least in our neighborhood of the universe -- became evident only in the late 19th and 20th centuries.

Accordingly, after Einstein's discovery that energy has a mass equivalence, the "zero-point-at-infinite-separation" convention for gravitation adopted in the distant past inevitably produces unphysical results -- "negative" energy implies negative gravitational mass exerting a negative gravitational "field", which has never been observed. Ultimately, then, this convention masks the fact that this "negative" potential energy must exert a positive gravitational pull if mass/energy is to be conserved. The existing flawed approach allows the possibly immense amounts of positive potential energy necessary to produce an arrangement of objects in space -- potential energy that can be visualized as an ideal fuel that can "burn" without leaving a trace other than

---

22 See J. Jewett, Jr. and R. Serway, <u>Physics for Scientists and Engineers</u>, 6th Ed., Section 13.6 with special emphasis on page at 404, especially Figure 13.13 (Thompson Brooks/Cole 2004).

23 See J. Jewett, Jr. and R. Serway, <u>Physics for Scientists and Engineers</u>, 6th Ed., Section 13.6 with special emphasis on page at 404 (Thompson Brooks/Cole 2004).

kinetic energy -- to be made to "disappear" by expending this energy to do work moving the objects away from each other. The fuel does work against the gravitational pull of the traditional matter (and against its own sometimes substantial gravitational pull) yet, under the present approach, merely reduces the negative amount of passive potential energy in a system. This cannot be real[24].

Although this point is simple, several examples are provided because the failure of the existing zero-point convention is so critical to understanding why the present gravitational framework is not valid as applied to large-scale systems. Accordingly, let us return to the example introduced above of two neutrons infinitely separated from each other.

---

24 To set up a relevant analogy, consider each pair of gravitating objects in the universe as a container for energy storage (energy is stored by moving the objects away from each other until the objects are infinitely separated). Pairs of "small" objects such as the Earth and the Moon represent "containers" with very limited capacities. Pairs of stars represent larger containers and pairs of galaxies represent containers of substantial size. Now consider three receptacles for fluids, a shot glass, a quart container and a 55 gallon drum. If we put an ice cube in each of the containers (as an analog for a "test" "mass" in a gravitational system) and fill each to the point where the cube is about to spill out, each container would contain vastly different amounts of fluid. If we assigned the quantity "zero" to the amount of fluid in each full container and defined the empty shot glass as containing a negative fluid ounce, the empty quart container as containing a negative quart and the empty 55 gallon drum as containing negative 55 gallons of fluid, we would have an approach to containers that mirrors the current approach to gravitating systems. Notwithstanding such an approach, who would rise to defend the assertion that the mass of the 55 gallon drum would be the same whether it really is empty or has but a single gallon in it or is filled to the brim? Can we, with a straight face, assert that the empty container contains - 55 gallons, the container with 1 gallon in it contains - 54 gallons while the filled-to-the-brim container actually contains none at all and thus the only mass we have to consider in lifting each container is the empty container's mass added to the mass of the ice cube? Certainly not.

By virtue of the existing zero-point convention, this arrangement contains no potential energy and, therefore, the mass of the system is simply the sum of the rest masses of the two neutrons. As we saw above, however, this system collapses into one containing the same two neutrons with presumably the same rest masses and also -1.04 x 10^{-66} Kilograms of potential energy[25]. In reality, of course, it is when the two neutrons are at infinite separation that they are in the highest energy arrangement of two such objects. The "negative" mass/energy figure at infinite separation developed above simply represents the positive energy that would be required to separate the two neutrons from each other if it is assumed that this potential energy quantity is too small to matter in the calculation. To be sure, the fact that the potential energy in the indicated system is so many orders of magnitude less than the mass of two neutrons makes clear that we can, indeed, ignore

---

25 The international standard of mass was, at least at one time, the (positive) kilogram based on a cylinder located at Sevres, France composed of platinum and iridium. See W.G.V. Rosser, An Introduction to the Theory of Relativity, Butterworth & Co. (Publishers) LTD. (1964) at page 3. Where in the universe would one go to find and hold in one's hand a negative kilogram? Would such a kilogram nullify the gravitational influence of the platinum-iridium standard if the two were brought together? The answers must be "nowhere" and "never". Science currently is of the view (and it is a view shared by the author) that it is impossible to shield objects from the gravitational influence of other objects and thus that there can be no negative kilograms anywhere in the universe. After the publication of Einstein's Special Relativity paper in 1905, it became impossible to justify negative Joules as well, since, in Special Relativity, all energy has mass and all mass is latent energy. Nevertheless, the practice of incorporating negative energy figures into calculations was so ingrained in the physics community that no one in the past hundred years or so seems to have given it a second thought. It is time, of course, to do so. The use of negative energy quantities should always be with a warning that the conclusions reached will only be valid on systems where the gravitational influence of energy can be ignored. To incorporate negative energy figures into a gravitational analysis of large systems will never work.

this potential energy so long as we are dealing with the basic building blocks of matter and modest accumulations of those building blocks. The Newtonian approach to the Solar System would have failed long ago were this not the case.

Consider, however, that astronomy currently accepts as valid the assumption that there are aggregations of neutrons in very dense clusters -- objects that are commonly referred to as "neutron stars". The escape velocity of such a star is reported to be on the order of one-half of the speed of light[26]. Because the escape velocity approaches a relativistic figure, we will use the different approach suggested in footnote 14, above, for inferring the mass equivalence of the kinetic energy and potential energy of an object that starts adjacent to our neutron star with just slightly less than escape velocity, moves away toward infinity and returns[27]. In this analysis, we will use the formula:

---

26 See M. Begelman and M. Rees, Gravity's Fatal Attraction, at page 46, Scientific American Library (Distributed by W.H. Freeman and Company) (1996).

27 Note that, as an alternative derivation of the relevant principle, consider the process of doubling the number of neutrons in our original system. If we analyze a system that consists of two pair of neutrons with each pair an infinite distance from the other pair, the potential energy in the system should be 4 times greater than our original system -- the gravitational force then operating, which is the product of the two mass quantities involved, would now be 4 times greater but the distance through which this more powerful force operates would remain virtually unchanged. Further, if we double the number of neutrons again to 8 -- two sets of 4 -- then the potential energy in the system should now be 16 times greater. Doubling again to 16 neutrons -- two sets of 8 -- results in a system with 64 times the potential energy. Further doublings of the number of neutrons floods the system with exponentially greater quantities of potential energy. Indeed, because this potential energy is a function of the square of the number of neutrons, we should be able to make this potential energy become the dominant form of mass in our system. This notwithstanding, however, our zero-point convention irrationally assigns the quantity "zero" to this ever increasing energy (and therefore mass) component.

$$m(v) \quad = \quad \frac{m_o}{(1-v^2/c^2)^{\frac{1}{2}}}$$

Applying this formula and assuming a moving object starting at the surface of the neutron star with just less than escape velocity -- i.e., with a velocity only just less than 50 percent of the speed of light -- then the moving object will have a mass of roughly 115 percent of its rest mass when viewed from the center-of-mass of the relevant system, as shown by the following computation:

$$m(v_{Esc}) \quad = \quad \frac{m_o}{[1-(.5c)^2/c^2)]^{\frac{1}{2}}}$$

$$m(v_{Esc}) \quad = \quad \frac{m_o}{[1-(.25)]^{\frac{1}{2}}} \quad = \quad \frac{m_o}{(.75)^{\frac{1}{2}}} \quad = \quad \frac{m_o}{0.866025}$$

Therefore,

$$m(v_{Esc}) \quad = \quad 1.154701\,(m_o)$$

Accordingly, the mass equivalence of the potential energy in the system that results when the moving object almost reaches the point of infinite distance (and begins to fall back) is roughly 15 percent of the rest mass of the distant object. This, to be sure, is just an approxima-tion because we have ignored the impact of this potential energy on the behavior of the almost escaping object and the mass equivalence of this potential energy is not really so small that it can be totally ignored -- it is only roughly a single order of magnitude less than the relevant object's rest mass. Accordingly, for the object to almost reach infinite separation, it would have had to begin its journey with still more kinetic energy than we have estimated, this additional kinetic energy would increase the overall mass in our system requiring still more initial kinetic energy to represent "escape" velocity, this additional energy would in-crease the need for initial kinetic energy still further and so forth in an infinite series that converges likely not too far from our approximation[28]. All this notwithstanding, the existing approach to gravitation blithely ignores this potential energy altogether. Those in the field have simply defined (by agreement among themselves) the system that consists of a neutron star and an infinitely distant object as containing no gravitational potential energy. It should be obvious at this point, however, that a system that consists of two neutron stars at infinite separation contains a quantity of gravitational potential energy that cannot really be ignored -- no sane observer would insert himself or herself between two such objects on a collision course.

Moreover, neutron stars are hardly the most exotic denizens of the universe according to existing astronomical thought -- such thought currently populates the universe with all manner of objects and systems that are far more massive. Among these are the now ubiquitous "black" "holes" that astronomers appear to be "finding" virtually everywhere. These objects have escape velocities that, incredibly, exceed the speed of light. We will have more to say regarding these fantastic objects later but, at this point, let us close this section with a series of examples involving hypothetical objects that are more massive than neutron stars -- objects with escape velocities that are more and more relativistic -- but that are not "black" "holes"[29].

---

28 The mathematics are similar to the math required to produce a gross income figure for someone who has an after-tax target salary. One starts with the target salary, one calculates the tax required, one adds that to the salary, one calculates the additional tax required on the additional tax payment and so on. The numbers will ultimately converge.

29 Newtonian mechanics developed a formula for relating the mass of an object with its escape velocity. That formula is:

$$V_{Escape} = (2GM/R)^{1/2}$$

Where G is the gravitational constant, M is the mass of the object and R is its radius. See J. Jewett, Jr. and R. Serway, Physics for Scientists and Engineers, 6th Ed., Section 13.7 with special emphasis on page 408 (Thompson Brooks/Cole 2004) While this formula may ultimately need revision in light of the ideas expressed in this work, it seems self-evident

Thus, imagine a massive object with an escape velocity of .95 percent of the speed of light. Take a standard "test" particle and launch the test particle radially away from the reference object at a speed of just less than .95 percent of the speed of light.[30] Again using the formula from the second of the concepts above, an individual in a reference frame at rest with respect to the central body (or more properly at rest with respect to the center-of-mass of the system) would initially measure the inertial and gravitational mass of the launched body at 3.20256 times the launched body's rest mass. As the launched object rises in the reference object's gravitational "field", however, its velocity declines until finally, just before infinite separation, it comes to a stop (because it did not quite have escape velocity at the point of departure). At this point, the visible masses in this system are the reference object and the test particle which are both at rest in the center-of-mass system we originally chose. For mass/energy to be conserved, however, the gravitational "field" in the space outside the two bodies and located between the reference object and test particle must now contain a mass

---

that, for a given radius, the greater the mass of the object, the higher its escape velocity.

30 This example is obviously imperfect because the author has ignored the recoil of the reference object. The author has done so for simplicity of illustration. To avoid the problem, it is certainly possible to launch two identical objects at identical velocities from opposite sides of the reference object. The mathematics will be more complicated (because each test particle would exert a gravitational pull on the other) but the qualitative result would not change materially. Similarly, the example is imperfect in the sense that the energy necessary to launch the test particle could not just "appear" as the particle was launched. This energy would have to be present already in the system (as chemical energy or nuclear energy or the energy of matter and antimatter awaiting annihilation, for example), and this stored potential energy would necessarily exert its own gravitational pull as well. Conservation of mass/energy necessarily requires that the system's gravitational pull never changes absent the passage of mass or energy through the system boundary.

distribution with an aggregate magnitude of roughly 2.20256 times the mass of the test object.

A more massive reference object -- an object with an escape velocity of .999 percent of the speed of light -- and a faster but otherwise identical test particle -- moving at just less than .999 percent of the speed of light -- will ultimately generate, after the test particle comes to a stop, a potential-energy-based gravitational "field" in the space outside the two bodies and located between them that must now have a gravitational mass of roughly 21.36627 times the mass of the test particle. Further increases of the escape velocity of the reference object or of the mass of the test particle or of both would still further increase the mass of the gravitational potential energy stored in the system.

Indeed, if one considers a reference object with an escape velocity of .9999999999 percent of the speed of light (obviously a massive object but not quite a "black" "hole") and our standard "test" mass and if this test object is ejected at just less than .999999999 of the speed of light, then, when this object comes to a stop, there would exist a potential-energy-based gravitational "field" in the space outside the two bodies and located between them that would represent roughly 2.236×10^4 times the mass of our "test" object.

As noted above, of course, the conventional use of a negative sign for this potential energy makes this enormous store of energy (and therefore gravitational mass) disappear. A simple convention as to the sign of a scalar energy quantity, however, cannot have such significance. If this arbitrary convention is reversed and the gravitational potential energy when an object is at an infinite distance from a gravitating body is assigned a positive value equal to the positive value of the mass associated with the kinetic energy of the object just before its impact with the gravitating body, the problem with "Dark" "Matter" may very well be solved. Dark Matter is the potential energy stored in gravitational "fields" and the significance of this form of energy as an independent source of gravitational pull has simply been obscured by the centuries-old assignment of a negative sign to it.

d. The Problem with the Negative Sign in the Classical Gravitational Potential Convention

We have previously used two distinct methods for assessing the potential energy located in a gravitating system that includes a dominant object. One method, the classical one, assigned the potential energy in a gravitating system a negative figure calculated using the following formula:

$$(\text{Potential Energy})_{\text{Gravitational}} = - \frac{G(M_1)\,(M_2)}{(r_{1,2})}$$

The second fixes the potential energy in such a system by using the fiction that the less massive object started adjacent to the dominant one and reached its current location by being ejected outward at the velocity needed to reach the relevant location. We have then used the mechanics of Special Relativity and specifically the second formula in 5.a., above to assign a positive mass/energy value to the potential energy contained in this system.

As a series of exercises -- the significance of which will become apparent when the exercises are complete -- let us use these two different methods to determine the potential energy associated with the two most massive components of our Solar System, the sun and the planet Jupiter. We will first use the classical formula to determine the potential energy associated with Jupiter at its mean orbital radius and then we will use the same formula to determine the potential energy associated with the planet Jupiter under the simplifying assumption that it is a compact "point" mass located at the sun's surface[31]. After the

---

31 By treating Jupiter as a "point" mass, we avoid having to deal with complexities associated with its radius.

second of these two computations, we will use the second method to compute the relativistic mass enhancement of the compacted planet Jupiter under the assumption that it is launched from the surface of the sun with escape velocity. The author believes that the results of the later two comparisons is quite revealing.

In our first calculation, we will take the mass of the sun to be 1.991×10^{30}Kg exactly; the mass of the planet Jupiter to be 1.90×10^{27}Kg exactly; and the average distance between the sun and the planet Jupiter to be 7.78×10^{11}m exactly[32]. Inserting these in the relevant formula, we find:

$$PE_{Grav. - Jupiter} = - \frac{6.673 \times 10^{-11}Nm^2/Kg^2 \, (1.991 \times 10^{30}Kg) \, (1.90 \times 10^{27}Kg)}{7.78 \times 10^{11}m}$$

$$PE_{Grav. - Jupiter} = - \frac{25.243 \times 10^{46}Nm^2}{7.78 \times 10^{11}m} = - 3.245 \times 10^{35} \text{Joules}$$

This quantity of gravitational potential energy has the following mass equivalence:

$$PE_{Grav. - Jupiter} = - \frac{3.245 \times 10^{35}\text{Joules}}{(2.997\,924\,58 \times 10^{8} \, m/sec)^2} = - \frac{3.245 \times 10^{35}\text{Joules}}{8.987\,551\,79 \times 10^{16} \, m \,^2/sec^2}$$

---

32 The planetary data used can be found in J. Jewett, Jr. and R. Serway, <u>Physics for Scientists and Engineers</u>, 6th Ed., at the inside front cover "Solar System Data" table and also in Table 13.2 on page 399 (Thompson Brooks/Cole 2004). We have applied the descriptor "exactly" to each quantity from these tables so that we can ignore the complexity associated with significant figures. This is necessary because the relativistic mass enhancement of a Jupiter-like object moving at the sun's escape velocity is many orders of magnitude less than Jupiter's rest mass and thus would be obscured by the amount of error in the original measurement of the planet's mass. Since we are not here attempting a true measurement, but, instead are comparing qualitative effects, we should not need to consider whether we could really test our calculations experimentally.

$$PE_{Grav. - Jupiter} \quad = \quad - 3.611 \times 10^{18} Kg$$

Were the planet Jupiter to be found just touching the surface of the sun -- the lowest energy position of these two bodies -- and were it treated as a simple point mass, then the quantity of gravitational potential energy in this arrangement would be calculated as follows:

$$PE_{Grav. - Jupiter} = - \frac{6.673 \times 10^{-11} Nm^2/Kg^2 \, (1.991 \times 10^{30} Kg) \, (1.90 \times 10^{27} Kg)}{6.96 \times 10^8 m}$$

$$PE_{Grav. - Jupiter} = - \frac{25.243 \times 10^{46} Nm^2}{6.96 \times 10^8 m} = - 3.627 \times 10^{38} Joules$$

This quantity of gravitational potential energy has the following mass equivalence:

$$PE_{Grav. - Jupiter} = - \frac{3.627 \times 10^{38} \, Joules}{(2.997 \, 924 \, 58 \times 10^8 \, m/sec)^2} = - \frac{3.627 \times 10^{38} \, Joules}{8.987 \, 551 \, 79 \times 10^{16} \, m^2/sec^2}$$

$$PE_{Grav. - Jupiter} = - 4.036 \times 10^{21} Kg$$

This quantity of gravitational potential energy has a mass equivalence, albeit a mass equivalence that is negative, that is roughly 28 percent of the mass of the "planet" Pluto[33].

In contrast, if we start with the planet Jupiter (treated as a point mass) just adjacent to the sun and launch the planet radially away at the sun's escape velocity[34], then, using the second of the formula in

---

33 As indicated in a prior footnote, the data used here can be found in See J. Jewett, Jr. and R. Serway, Physics for Scientists and Engineers, 6th Ed., at the inside front cover "Solar System Data" table (Thompson Brooks/Cole 2004).

34 The escape velocity of the sun can be found in J. Jewett, Jr. and R. Serway, Physics for Scientists and Engineers, 6th Ed., in table 13.3 at page 408 (Thompson Brooks/Cole 2004).

5.a., above, the relativistic enhanced mass of the planet would be as follows:

$$M(v_{Jupiter-esc-Sun}) = \frac{(1.90 \times 10^{27}Kg)}{[1 - (6.18 \times 10^5 \ m/sec^2/2.997\ 924\ 58 \times 10^8 \ m/sec)^2]^{1/2}}$$

$$M(v_{Jupiter-esc-Sun}) = \frac{(1.90 \times 10^{27}Kg)}{[1 - (38.19/8.987\ 551\ 79 \times 10^6)]^{1/2}}$$

$$M(v_{Jupiter-esc-Sun}) = \frac{(1.90 \times 10^{27}Kg)}{[1 - 4.250 \times 10^{-6}]^{1/2}}$$

$$M(v_{Jupiter-esc-Sun}) = \frac{(1.90 \times 10^{27}Kg)}{(.99999575)^{1/2}} = \frac{(1.90 \times 10^{27}Kg)}{.999997875} = 1.900004038 \times 10^{27}Kg$$

Therefore, if we treat the planet Jupiter as a point mass and if we assume that it starts adjacent to the sun moving radially outward with escape velocity, then its mass, viewed from the center-of-mass of the system, would be $4.038 \times 10^{21}Kg$ greater than its rest mass. Initially, this "mass" is associated with the escape kinetic energy and is localized with the moving object.

Note that we obtained *virtually the same numerical value* for the potential energy of a stationary point-mass Jupiter located at the sun's radius using the traditional Newtonian potential energy formula *except that, significantly, the figure generated by the traditional formula was a negative number.*

Imagine, then, that we begin with our point-mass Jupiter at infinite separation from the sun. Traditionally, this system has only two components, the sun and the point object. The mass of this system in current thought, therefore, would be the sum of the rest masses of these two objects only. Accordingly, using current concepts, we would construct a gravitational "field" for this system by selecting each point in space and determining the vector sum of the gravitational forces caused by (1) Jupiter's rest mass acting at the relevant point's current distance from Jupiter; and (2) the Sun's rest mass acting at the relevant point's current distance from the Sun[35].

As the compacted planet falls in toward the sun, its speed increases until, just before impact, it would have roughly escape velocity. Viewing this system from its center-of-mass, the overall system mass the instant before impact would be the mass of the sun plus the mass of the planet plus the mass equivalence of the final kinetic energy figure. If we assume an inelastic collision, then the mass of the resulting object as measured in the center-of-mass system would be equal to the sum of these three discrete components.[36] The traditional gravitational "field" of this object, then, would be the vector sums of Jupiter's rest mass field plus the Sun's rest mass field plus the field associated with the kinetic energy that Jupiter accumulated as it fell. If, in contrast, we assume an elastic collision, our point-mass Jupiter would rebound with almost escape velocity. When our point Jupiter almost reaches infinite separation, however, under a traditional analysis, the mass of the system would again be just the rest mass of the sun plus the rest mass of the planet. Using current concepts, we could repeat this cycle

---

35 See, See, D. Tilley, University Physics for Science and Engineering, Cummings Publishing Company, Inc., Menlo Park, California (1976), at section 14.1 and especially figure 14-2 (where the text reads: The superposition principle. At point P, mass M_1 alone would produce field g_1; M_2 alone would produce g_2. When both sources are present the field is the vector sum of g_1 and g_2). See also, P. Tipler, Physics, Worth Publishers, Inc. (1976), starting under section 16-3 at page 398.

36 See, French, A.P., Special Relativity, W.W. Norton & Company, Inc. (1968), pages 172 to 175; and See W.G.V. Rosser, An Introduction to the Theory of Relativity, Butterworth & Co. (Publishers) LTD. (1964) at section 5.7 beginning on page 217. Each of these discussions details the result of a collision between two objects as perceived using the mechanics required by Special Relativity. This analysis will be considered over and over again in the discussion that follows.

indefinitely, forever "creating" and then "destroying" the mass associated with the kinetic energy that appears and then disappears from the system.

In reality, of course, the mass/energy of this system should not vary and one of our assumed concepts above is that it does not. In a proper analysis, then, we expect that, when viewed from the center-of-mass of the relevant system, the system mass/energy is the rest mass of Jupiter plus the rest mass of the sun plus the unvarying positive mass associated with: (1) the kinetic energy in the system the instant before impact; or (2) the potential energy in the system at infinite separation; or (3) the sum of the remaining potential energy and the then existing kinetic energy at any distance between infinite separation and contact. Having mistakenly, by historical accident, assigned the quantity of the potential energy at infinite separation a negative value, existing thought makes the third component of this system appear to wink in and wink out -- at infinite separation, it "goes dark" and becomes "Dark" "Matter" in our system only to "reappear" to the full extent at impact[37]. Of course, if our system contemplates the object's indefinite stay at a particular separation -- if our object is in an orbit within the system -- then the potential energy in the system remains "dark" indefinitely. Objects under the gravitational influence of this system would behave differently than we expect based on existing concepts of gravitation because those objects would be under the influence not only of the component objects in the system but also the influence of the system potential energy. We would have a "dark" "matter" problem -- one of the very problems that physics (not surprisingly in this author's view) faces today. If the

---

37 Note well the explosive growth of "mass" in the collapse of all structures under the current analysis and a frequent statement that every object, even the humble earth, has a radius that, were the existing mass to be compacted to the relevant radius, the object would become inescapable -- a "black" "hole". For the earth, that radius is alleged to be 0.9 centimeters and for the sun, that radius is alleged to be 3 kilometers. See Carmeli, Moshe, Classical Fields, General Relativity and Gauge Theory. World Scientific Publishing Co. Pte. Ltd. (2001)

discussion above is accurate, then, trying to find additional objects in the night sky in order to "solve" this problem will never be successful. The more objects that are found, the more explosive will be the growth of system potential energy and the more "dark" objects that will need to be found. If the author is right, then, no one will ever find the "dark" "matter" for which physics is currently searching.

e.   Further Discussion of Problem Concepts in the Foundations of Gravitational Theory

Several final points remain to be made in regard to the series of examples above drawn from our Solar System. First, it is clear that the objects that we have analyzed are of sufficiently modest mass that the potential energy implicit in their arrangement has little discernable impact on the overall behavior of the system. We have analyzed the interaction of the two most massive components of the system and found that the mass equivalence of the system potential energy implied by their separation is less than the mass of the least massive of the planets.[38] Our candidate for "Dark" "Matter", then, is sufficiently ineffective within the Solar System that it would cause only a minor deviation from the behavior predicted with Newtonian concepts. Accordingly, our approach is on a similar plane with General Relativity which is largely ignored in any analysis of the Solar System except as a perturbation to Newtonian concepts that is called on only to explain the precession of the perihelion of Mercury's orbit and little more.

---

38 An estimate of the quantity of potential energy associated with the separation between the planet Jupiter and the sun would be the absolute value of the difference between the potential energy at closest approach for the point-mass planet, which was - 4.03 x 10^{21}Kg, less the potential energy figure at the average orbital radius of the planet, - 3.61 x 10^{18}Kg. As noted above, this represents less than 28 percent of the mass of the planet Pluto.

Of course, in addition to fixing a quantity to the mass-equivalence of the system potential energy we are also going to have to suggest the location from which its influence is exerted. Moreover, we are going to have to deal with the fact that assigning this quantity of gravitational potential energy to the "empty space" between objects is inherently problematic based on existing physical concepts. If these objects are orbiting each other, then the empty space that must "contain" this mass is constantly changing throughout a planet's orbital path -- if we could decelerate the planet Jupiter in its orbit and thereby cause it to collapse into the Sun, the mass of our potential energy would have to be located along different paths at different times during Jupiter's normal cycle. Moreover, the sun and Solar System are, themselves, in orbit around a galactic center so that, even when the planet Jupiter has reached the "same" point in two successive orbits, it is not at the "same" point relative to the center-of-mass of our galaxy. Accordingly, it will be difficult to reconcile our assumed principle that mass/energy is conserved and our assertion that the potential energy in gravitating systems acts gravitationally. We will discuss this problem and a possible solution to it much further below.

At this point, however, the author would note that all "objects" that have been discovered so far are really just aggregations of complex electro-magnetic and strong force operative fields -- most of the building blocks of such "objects" we call "atoms" and we understand them far better than anyone understood them at the time Albert Einstein developed his relativity theory. The author would further assert that the appearance of "inertia" and therefore of "mass" for these objects is entirely the result of (1) the energy content of these fields as suggested in footnote 21, above, and (2) the entirely local process by which the energy in such fields is conserved when these fields are forced to "change" -- changes at a point over time are always matched by compensating spacial changes so long as a consistent frame of reference *and* the Lorentz transformation equations are used within frames of "objects" moving with respect to each other. Accordingly, if there is a rule that requires that energy be conserved in all local electro-magnetic changes and manipulations -- and there is -- and if there is a similar rule that requires energy be conserved in local strong force manipulations, then all permissible manipulations of "objects" should result in overall local energy conservation.

In this regard, consider that a stationary charged particle implies the existence of an electric field that is radially inward throughout all of space and regardless of any other frame of reference in which the field is analyzed. This field is spherically symmetrical, however, only in its rest frame[39]. A second charged particle does not directly interact with the first "particle" over a finite distance. Instead, the second charged particle interacts locally with the different electric and magnetic components of the first charged particle's pre-existing field as measured in the second particle's "rest" frame. Thus, it is said that the influence of the first particle on the second arises from the "retarded position" (not the current position) of the first. As a result, although the second charged particle is influenced by what that second particle would anticipate to be the "current" position of the first, this is only an illusion -- a consequence of the application of the Lorentz transformations required to allow the second charge particle to compute the magnitude of the fields of the first in its own rest frame and at its own rest position. It is the local

---

39 See E. Purcell, Electricity and Magnetism, (McGraw-Hill Book Company 1965) ¶5.6 beginning at page 158 and with special emphasis on page 160. See also, French, A.P., Special Relativity, W.W. Norton & Company, Inc. (1968), pages 237 to 250, especially at 242 to 243 and W.G.V. Rosser, An Introduction to the Theory of Relativity, Butterworth & Co. (Publishers) LTD. (1964) at section 7.4 beginning on page 290 with special emphasis on the discussion at pages 291-292.

field components and local energy content implicit in the construction of the distant particle's field that govern the first particle's behavior at this local point[40].

Returning then to the collapse of the planet Jupiter into the Sun, it becomes clear that, while physicists have, for generations, imagined the process of bringing the planet Jupiter to a halt in its orbit and watching it fall into the Sun, we know (as, in fact, they knew) that this is not physically possible. One could only split the existing planet into pieces and send a portion into the sun at the expense of sending another portion somewhere else entirely[41]. If we leave aside for a moment the prospect of tapping the energy in the nuclei of the atoms of Jupiter and consider only access to chemical energy to assist us in this process, the result would be that we would tap stored potential energy in the complex electro-magnetic fields of the existing components of Jupiter by creating -- through a "chemical" reaction -- a local electric or magnetic field that acts on the local electric and magnetic components of other portions of the system. These local field modifications then result in changes that we interpret as "objects" accelerating away from each other. These net accelerations, operating over time, convert some of the electro-magnetic potential energy stored in the sub-components of

---

40 To be sure, the first particle may never make it from the retarded position to the "current" position -- it may have experienced an acceleration before reaching the location where the second particle would expect it to be. On the other hand, it would take the impact of some force to prevent the first particle from reaching the anticipated "current" position, and, of course, if we are assuming that all forces are ultimately electromagnetic in origin or consequence, then the impact of the electromagnetic field sources that have influenced the first particle and prevented its arrival at the "current" position would also have an impact on the second and this impact would also be local as perceived in the second particle's rest frame.

41 See Valens, E. G., The Attractive Universe: Gravity and the Shape of Space, The World Publishing Company (1969), for comments on a hypothetical collapse of the planet earth into the sun and the complications involved.

Jupiter into the kinetic energy of the electro-magnetic sub-components performing the observed motions. Every step in the process is caused by a local electro-magnetic interaction and the requirement that energy be conserved in all frames of reference (that are not accelerating) during the course of such interactions and the further requirement that changes in the locations of all field sources be communicated at a finite speed through a process that also conforms to local energy conservation insures that the overall process happens just the way we expect based on Newton's understanding of gravitation, mechanics, and energy conservation.

Gravity, therefore, is not a separate force operating between distant "masses" but, instead, is an apparent behavior of voids in the system of electro-magnetic fields because of the constraints on changes to such fields -- the constraints of charge conservation and the invariance of Maxwell's equations regardless of the "velocity" of the hypothetical observer who could be experiencing these fields both leading to interactions between the "current" positions (created at the time of the retarded positions) of all charged objects in the universe. Accordingly, as noted above, there are no "neutral" objects in the universe, per se. There are just local conditions that may superimpose attractive and repulsive fields of equal strength and direction at a particular point. The local cancellation of these fields cannot extinguish the fields of the source particles everywhere if there ever was a time at which the source particles were separated from each other. We will come back to these ideas later but must move on now with the destruction of the old flawed system before attempting to build anew lest any flaws in the new approach provide an opening for the proponents of the old to cling to what must be discarded[42].

---

42 To foreshadow what is to come, consider the discussion in Tolman, Richard C., Relativity Thermodynamics and Cosmology, Oxford at the Claredon Press (1958), beginning at Section 39, page 84 and continuing on through Section 47 ending on page 100. If one simply comes to the realization that there is no "mechanical" behavior that is not electrical or magnetic in character -- i.e. that our mechanical concepts including heat, pressure, the work done by expanding gasses such as steam,

Suffice it to say, however, that the author is not associating "mass" with "empty" space. Instead, he is recognizing that space cannot truly be "empty." Instead, it must be everywhere filled with the superposition of all of the electric and magnetic fields of each and every discrete charged proton, electron and other charged elementary particle in the universe -- even the field of a single, lowly electron classically stretches to infinity in all directions and every volume of the field of that single, lowly electron has a defined energy and therefore mass content at each point short of infinity and outside its radius. To be sure, one may bring a positive charge from some other place to "neutralize" the field of this electron but the fact that these two charges now cancel in their immediate vicinity does not mean that the complete history of the series of motions which brought these charges together has been erased throughout the universe. Quite the opposite is true. The history of these motions is propagating through the universe in conformity with Maxwell's equations.

Of course, in free space there are no protons, electrons or other "particles" and thus there are no consequences to propagation of these changes that we would perceive and interpret as "objects" behaving in any particular manner. It is only when these free space changes reach what we have traditionally associated with "matter" -- the boundaries between the interior and exterior of an electron or proton or the like -- that we witness phenomena that we identify as objects with "mass" accelerating or decelerating or otherwise being affected. This notwithstanding, the free-space changes must be as real as the changes to

---

etc., are simply the macroscopic consequences of the innumerable collisions between the basic electrical structures we call atoms and their combinations -- then it becomes clear that attempts to combine the results of "mechanical" and "electromagnetic" actions as suggested in Section 45 of Tolman's work are a mistake. Take the mechanical discussion out and simply read and consider how electrical systems work and couple this with the established superposition concept that is appropriate in electrodynamics and one likely will have a sound foundation for the future understanding of the real way the universe works.

what have traditionally been described as "objects." Indeed, every text the author has read associates an energy content -- given by the Poynting vector[43] -- with every volume element of the space through which electro-magnetic radiation is moving.

It takes very little imagination, then, to visualize our Sun, which existing thought suggests has been emitting electromagnetic radiation over a broad spectrum and in all directions for billions of years, as the center of an extensive system of energy -- and therefore mass -- independent of the body of the star itself[44]. The further one goes from the center-of-mass of the Sun, the greater the significance of the mass of

---

43 The Poynting vector, **S** is related to the electric intensity **E** and the magnetizing force **H**, by the relation **S** = **E** X **H** with this interpreted as the amount of energy crossing a unit area perpendicular to the vector **S** per second. <u>See</u> W.G.V. Rosser, <u>An Introduction to the Theory of Relativity</u>, Butterworth & Co. (Publishers) LTD. (1964) at section 5.8.3 beginning on page 222 with special emphasis on the discussion at pages 223. <u>See also</u>, W.K.H. Panofsky and M. Phillips, <u>Classical Electricity and Magnetism</u>, Addison-Wesley Printing Company, Inc. (1956) at §10-6, especially at 163.

44 An internet source indicates that the sun releases 3.8310×10^{26} Watts (383 yottawatts) of energy per second. The mass equivalence of this energy quantity is 4.2626×10^9 kilograms. Accordingly, the mass equivalence of the energy output of the sun in a year is (4.2626×10^9 killograms/second) (60 seconds/minute)(60 minutes/hour)(24 hours/ day)(365 days/year) or 1.3443×10^{17} kilograms per year. If the sun has been radiating for a billion (10^8) years, then, in the space between the sun and an object a billion light years away, there is a quantity of mass equal to 1.3443×10^{25} Kilograms -- roughly the mass of the planet Uranus -- tied up in this radiation. At the start of this billion year period, the sun had more Hydrogen and less Helium than at the end of the period and was therefore more massive as a discrete body. At the end of the period, the mass of the discrete body has been reduced to fund the surrounding sphere of radiated energy. To an object beyond a billion light years from the Sun, however, the change in the form of the mass/ energy from the body of the sun to radiation now moving away from it cannot make a difference.

this radiant energy to the behavior of objects still further out. The mass equivalence of the radiation which has already passed a certain distance would have no influence on objects within just as a spherical shell of matter has no influence on objects inside the shell. The energy and therefore mass of this solar radiation, of course, can only be assigned to the "empty" space through which that solar radiation is passing.

Note also that the "collision" of two "objects" that are electrical in origin can only be made sensible if the collision is interpreted as involving the fields -- operating throughout all of space -- of which the objects are sources rather than the "objects" themselves.[45]

---

45 A complete discussion of such a collision can be found in W.G.V. Rosser, An Introduction to the Theory of Relativity, Butterworth & Co. (Publishers) LTD. (1964) at section 7.5 beginning on page 294 with special emphasis on the language from page 297. Mr. Rosser begins by noting that the forces between pairs of moving charges deviate from action and reaction by terms of the order of v^2/c^2. He continues by providing an interpretation of such deviations using classical electromagnetic theory. According to Rosser, citing works by Abraham and Becker and Page and Adams: "[i]t can be shown using classical electromagnetic theory that, if one associates a linear electromagnetic momentum density of $1/c^2$ $\mathbf{E}$ X $\mathbf{H}$ with the electromagnetic field, then the sum of the momentum of the matter and the momentum of the electromagnetic field in an isolated system of charges is constant with respect to time, though the law of action and reaction is no longer true for matter considered on its own". Mr. Rosser illustrates this interpretation with an example that consists of two charges, a charge q_2 moving along a coordinate axis to the right giving rise to fields $\mathbf{E}_2$ and $\mathbf{H}_2$ at a point in empty space and a charge q_3 to the right of q_2 and above the coordinate axis and moving to the left and producing fields $\mathbf{E}_3$ and $\mathbf{H}_3$ at the same point. According to Rosser, [t]he electromagnetic linear momentum per unit volume of this system totals: $1/c^2$ ($\mathbf{E}_2$ X $\mathbf{H}_2$ + $\mathbf{E}_2$ X $\mathbf{H}_3$ + $\mathbf{E}_3$ X $\mathbf{H}_2$ + $\mathbf{E}_3$ X $\mathbf{H}_3$) and "[t]he two terms $1/c^2$ $\mathbf{E}_2$ X $\mathbf{H}_2$ and $1/c^2$ $\mathbf{E}_3$ X $\mathbf{H}_3$ remain constant, if the velocities of the charges remain constant; [and] these terms are responsible for the electromagnetic masses of the particles when they are accelerated." In contrast, according to Rosser, "[t]he magnitudes of $\mathbf{E}_2$ and $\mathbf{H}_3$ vary at all points in space such that $1/c^2 \int \mathbf{E}_2$ X $\mathbf{H}_3$ dV, integrated over all of space

The reader may perceive in this discussion what is now heresy in physics -- the unabashed resurrection of the currently disfavored concept of an electromagnetic "ether." So be it. The author has never felt comfortable with the current fashion in physics and its denial of the obvious and willingness to embrace without discomfort mutually exclusive concepts[46]. Electro-magnetic waves are and always have been

---

varies with time and, [s]imilarly, $1/c^2 \int \mathbf{E_3} \times \mathbf{H_2} \, dV$ varies with time. Rosser concludes by noting that work by Page and Adams shows that the quantity "$d/dt \{\int (\mathbf{E_2} \times \mathbf{H_3} + \mathbf{E_3} \times \mathbf{H_2}) \, dV\}$ balances the differences between the forces between the charges illustrating how, according to classical electromagnetic theory, the sum of the linear momenta of matter plus the electromagnetic field is constant". Similar ideas can be found starting at section 10-5 on page 160 and ending at the conclusion of section 10-6 on page 164 of Classical Electricity and Magnetism by W.K.H. Panofsky and M. Phillips, Addison-Wesley Printing Company, Inc. (1956). Indeed, to the author, the discussion in Panofsky and Philips represents the cornerstone of a proper theory of mechanics -- "mechanics" as such is simply the study of that portion of electromagnetic fields that respond to the Lorentz force. Mechanics, therefore, is only a partial understanding of more complex electromagnetic phenomena and its continued separation from the larger body of electromagnetic theory is a relic of the historical development of our view of the universe. In summary, then, the discussions of Mr. Rosser and of Panofsky and Philips indicate that the concept of inertia is due to the energy conservation and induction properties of electro-magnetic phenomena. The author suspects as well that, if the analysis above were performed in a "center-of-mass frame" for the two charged particles, the constraint of energy conservation in this single frame would prevent either charged particle from moving on to their separate infinities. Further, the predicate for the example -- that there can be two charged particles moving at constant speed toward each other is unrealistic because the source particles would accelerate in response to the fields of each other regardless of the relative charges.

---

46 The following quotation from a relatively modern text, with an imbedded quotation from Maxwell, illustrates the mind set that is currently popular:

Now comes the really difficult question. Is electromagnetic radiation a thing? The direction of flow of radiation certainly does not fail to pass the test of invariance from frame to frame. Not only does the direction stay the same but its speed remains invariant from frame to frame. Why should we then resist calling it a thing? -- particularly when one of the important ideas of this century is that the energy E of electromagnetic radiation is quantized into packages (photons) that are proportional to the frequency v, $E = hv$, where h is Plank's constant. Is a photon a thing?

Maxwell struggled with this problem long before photons. He wrote:

> Now we are unable to conceive of the propagation of electric action in time except either as the flight of a material substance through space or as the propagation of a condition of motion or stress in a medium already existing in space.

In this statement he says *either* electric action itself is a thing, or there is a medium (a thing) all over space (the ether). Experiments to detect the ether have all failed, and Einstein has declared we should not speak about a thing that can never be detected. That would be metaphysics again. If ether is out, that would seem to leave the photon as a thing. But the photon is what is identified with electromagnetic waves (radiation). What about other electric actions such as the simple interaction between static charges? There would be just as much need for a thing to fly from one to the other here as when the charges are oscillating and producing what is called radiation. To explain this by the exchange of virtual photons, as is often done seems somewhat circular and unclear to say the least. That is why we have chosen not to treat electromagnetic fields as things. In this book fields are treated as constructs that help us to calculate the way that things (particles of matter) interact. Perhaps there are things flying around, but no one knows, and possibly we never will.

disturbances. To presuppose that they are disturbances in an all per-vasive medium is only natural and the coupling of the now mature un-derstanding of how "the" electromagnetic universe is perceived by ob-servers (who are themselves electromagnetic field constructs moving through this universe) as informed by an understanding of the Lorentz transformations, makes the "ether" far less mysterious. Indeed, the "real" ether -- a Euclidian 3-dimensional space of physical points with each point associated with the superposition of all of the evolving elec-tric and magnetic field intensities dictated by Maxwell's Equations at that particular point, whether that physical point is "empty" or within a system of what modern thought considers field sources -- sounds very much like the conception of electricity of one of the pioneering giants in this discipline, Michael Faraday[47]. Mr. Faraday, of course, was very

---

See, D.G. Ivey and J.N. Patterson Hume, Physics (In Two Volumes), Volume 2 (The Ronald Press Company New York 1974) at pages 10-11. The author, of course, agrees with Maxwell, we must pick one. The author picks the ideal of the propagation of a condition of motion or stress in a medium already existing in space as most compatible with what we now understand about electromagnetism.

47 To integrate and make sense of this evolving set of points, it is necessary, of course, for each observer to develop a conception of "time." Time to any specific observer, however, can only be measured by the counting of repetitive behaviors of electromagnetic constructs, and, because there is a single, integrated evolving universe, each observer's countings of repetitive behaviors and therefore each observer's time measurements are intimately related to the measurements of all others. The Lorentz Transformations provide the mechanism that must be used to translate each observer's distinct time and location measurements into a form that can be applied to the time and location measurements of others and, of course, Special Relativity makes clear that each observer's conception of his or her own time will affect his or her measurements of what is happening at each universal point. Thus, since the Lorentz Transformations link each observer's perception of spacial changes with that observer's perception of time, observers in relative motion cannot directly make measurements employing a single common time discoverable by review of each's local clock. This notwithstanding, if

concerned that the over-elaboration of mathematical constructs by the physicists of his day would lead physics astray. The author believes he was right.

f.   Indications that the Problem Concepts Above Remain Part of the Theory of General Relativity

At this point, it also would be well to compare the analysis above to a recent comprehensive analysis of a two body problem applying the concepts of General Relativity. Unfortunately, the author has not been able to find any such recent treatment. Every recent discussion of General Relativity the author has found has been based on analyses (complex, obtuse and filled with dubious assumptions) of single objects in the apparent (but not explicitly stated) belief that the fields of discrete objects can be superimposed on each other to develop a definitive approach to overall gravitating systems. This, of course, is what has traditionally

---

each were given the location of the center-of-mass of the system in which all were components, each could translate their measurements into the measurements that would be made in that center-of-mass frame and all should agree on those center-of-mass measurements and on the fact that the center-of-mass is, indeed, the center-of-mass. See Rindler, W., Essential Relativity, at Section 5.7 (Springer-Verlag 1977), To suggest an imperfect analogy as to the inter-relationship of measurements of time and space, consider an individual riding in a car on a roller-coaster. The roller-coaster represents a single, fixed physical sequence of movements but a whole range of physical experiences depending on how rapidly the car completes the entire ride cycle. If we are forced to use repeating physical experiences in developing an assessment of time -- if we can only count the numbers of ups and downs of the ride or some other physical process that is influenced by our motion as a measure of the duration of the ride -- then assessments of time and of space inevitably will be interrelated. This notwithstanding, there is only one roller-coaster just as there is only one universe and there is absolute significance to each rider's measurements.

been done when applying the Newtonian approach that has been criticized above including the discussion in footnote 6, above, which the author believes is compelling[48].

Although a comprehensive accessible analysis of a two body system is lacking, there is no shortage of language in a variety of texts that provides confirmation that the Newtonian and General Relativity approaches to gravitating systems are identical even though their approaches to the gravitational influence of discrete objects are quite different. Thus, in Bergmann's "Introduction to the Theory of Relativity[49] the following discussion of the construction of the gravitational "field" of a system of objects appears:

Preparation for a relativistic theory of gravitation.
Before we can hope to create a relativistic theory of gravitation,

---

48 This, of course, is also done with electrical systems -- when we construct the electric field direction and intensity at a particular "point" in space, we superimpose the fields of each charge in the universe at that point. The result is the aggregate field attributable to all charges at that location. Since charge is conserved regardless of state of motion, this superposition approach is appropriate and allowable. Mass, however, depends on relative motion and therefore it should not be possible to use the same approach for both electric and gravitational "fields". Observers in different frames of reference will assign different masses to each object in the universe thus preventing them from applying the same methodology to construct gravitational "fields" as to construct electric fields. In the same vein, energy stored in an electrical system acts only gravitationally and thus has no impact on an electrical analysis but energy stored in a gravitational system must act gravitationally and therefore must have an impact on a proper gravitational analysis. Proper physics, therefore, should not use identical approaches to electrical and gravitational systems and the fact that it does only confirms that the current understanding of gravitation must be deeply flawed.

49 See B. G. Bergmann, Introduction to the Theory of Relativity, Prentice Hall, Inc. (New York, 1947) at 154-155

we must first attempt to reformulate Newton's theory so that action at a distance is eliminated. This can be done fairly easily. The gravitational attraction of one body with the mass m by several other ones can be represented by the sum of the "gravitational potentials," (10.3), of these other bodies; this sum represents the potential energy U_m of the first body divided by its mass m. The force experienced by that body is the negative gradient of its potential energy,

$$f = m \operatorname{grad} G. \quad (10.5)$$

The gravitational potential depends on the positions of the other bodies. The contribution of every mass point is given by eq. (10.3)[50]. If we introduce a "gravitational field strength,"

$$g = - \operatorname{grad} G, \quad (10.6)$$

we find, just as in electrostatics, that the gravitational lines of force neither originate nor terminate outside of masses, and that, in a mass M, $4\pi\kappa\rho$ lines of force terminate. We conclude that the divergence of g is

$$\operatorname{div} g = - 4\pi\kappa\rho,$$

where ρ is the mass density. The potential G itself satisfies the equation:

$$\operatorname{div} \operatorname{grad} G \equiv \nabla^2 G = 4\pi\kappa\rho. \quad (10.7)$$

This equation, which was first formulated by Poisson, is, then, the classical equation of the gravitational field. Eqs. (10.5) and (10.7) together are completely equivalent to the equations of Newton's theory, which is based on action at a distance.

Poisson's equation, (10.7) is not Lorentz-invariant. Wherever ρ vanishes, it seems reasonable to assume that the three dimensional Laplacian operator ∇^2 has to be replaced by its four dimensional analogue, the operator:

$$\acute{\eta}_{\mu\nu}\partial^2/\partial x^\mu \partial x^\sigma = \partial^2/\partial t^2 - c^2\nabla^2$$

In the presence of matter, we must remember that the mass density ρ is not a scalar, but one component of the tensor $P^{\mu\nu}$.

---

50 This equation is: $G_M + \kappa M/r$ where G_M is the gravitational potential at a distance r from the mass point with a mass of M.

We face the alternative of either replacing ρ by the Lorentz-invariant scalar $\acute{\eta}_{\mu\upsilon}P^{\mu\upsilon}$ or replacing the nonrelativistic scalar G by a world tensor $G^{\mu\upsilon}$.

The author finds this discussion particularly important because Mr. Bergmann was a friend and confidant of Albert Einstein and, even more significantly, because Albert Einstein wrote the foreword to the book and therefore presumably would have corrected the discussion above if it were inconsistent in material respects with the concepts of General Relativity as he understood them. To the author, the discussion above leaves little doubt that General Relativity is a theory of objects and empty space, just like the Newtonian theory it purported to replace -- it retains the assumption that the gravitational potential at a given point can be associated with the superposition of gravitational potentials of discrete gravitating bodies and can be measured by use of a standard object. Thus, in General Relativity, there are only objects with mass values just like in the Roman numbering system there are only symbols with unvarying numerical values all as detailed in footnote 6., above. Because there is no place value concept in General Relativity, it cannot be "right." The complex mathematics of General Relativity and its references to "space-time curvature" only conceal the flaws in what the author and Albert Einstein agree is the requirement that potential energy stored in gravitational "fields" act as its own source of gravitational force in a proper understanding of gravitation and mechanics.[51]

---

51 See again, A. Einstein, The Principle of Relativity §16, toward the bottom, where the following language appears "It must be admitted that this introduction of the energy-tensor of matter is not justified by the Relativity postulate alone. For this reason, we have here deduced it from the requirement that the energy of the gravitational "field" shall act gravitatively in the same way as any other kind of energy."

The author would hasten to add that Bergmann's text is not alone in suggesting that General Relativity is exclusively an "object" value theory. Thus, in Frankel's Gravitational Curvature[52], the discussion in Chapters 3, 4 and 5 clearly indicates that, in General Relativity, gravitation is the result of bodies and not systems. The entire discussion there refers again to Poisson's Equation -- a long-standing mathematical relationship between the second spacial derivative of an energy function and the spacial distribution of "material' that "creates" the force with respect to which the energy function is defined. Similarly, according to Tollman:

> In accordance with the Newtonian theory of gravitation the action of gravity at any point in space at a given instant is determined by the location of the surrounding matter, and this general idea with suitable modifications must evidently be taken over into the relativistic theory of gravitation since the Newtonian theory is in any case an exceedingly close first approximation. In Newtonian theory the dependence of the gravitational potential ψ on the distribution of matter of density ρ is given by Poisson's equation:
>
> $$\partial^2\,\psi/\partial x^2 + \partial\psi^2/\partial y^2 + \partial^2\psi/\partial z^2 \;=\; 4\pi k\rho$$
>
> where k is the gravitational constant. In the relativistic theory of gravitation modifications will be necessary, in the first place since we shall need to calculate the ten components of the metrical tensor or gravitational potentials $g_{\mu v}$ instead of the single gravitational potential Ψ for Newtonian theory, and in the second place because the special theory of relativity has provided us with relations between mass, energy, and momentum which indicate that covariant expressions are to be obtained by making use of all ten components of the energy-momentum

---

52 See T. Frankel, Gravitational Curvature (An Introduction to Einstein's Theory), W. H. Freeman and Company (1979) at 27-55.

tensor $T_{\mu v}$ rather than singling out some single quantity which we could call *the* density of matter.

Our general aim, hence, will be to obtain a covariant equation connecting the $g_{\mu v}$ with the $T_{\mu v}$ which will be the analogue of Poisson's equation and which will lead to the same results as the Newtonian theory in first approximation. Before proceeding to the complete solution of this task, however, it will first be profitable to consider two special cases, that of the field corresponding to the special theory of relativity, and that of a field in empty space but in the neighborhood of gravitating bodies.

The general hypothesis that the metrical field is determined by the distribution of matter and energy may be call the principle of Mach.

See Tolman, Richard C., Relativity Thermodynamics and Cosmology, Oxford at the Claredon Press (1958), Section 75, page 184-185. See also Tolman, Richard C., Relativity Thermodynamics and Cosmology, Oxford at the Claredon Press (1958), Sections 77 and 78 discussing gravitational "fields" in "empty space" and in the presence of "matter." As the author has suggested above, however, there is no "empty" space and no objects of "matter" per se. There are simply concentrations of energy that appear intense when energy is stored in familiar electrical systems and diffuse when stored in familiar gravitational systems.

Given the obvious defects with the Newtonian approach critiqued above and literature quoted above that shows that the General Relativity approach is blind to these defects, how much confidence should we then have in General Relativity's approach? How, in fact, does the space-time curvature conception central to General Relativity[53] address

---

53 What, in fact, is space-time curvature? The literature is rife with references to it but the author remains without a clear conception of what it is and how it is alleged to operate. Indeed, to the author, it appears an incantation

the influence of gravitational potential energy on an overall gravitating system? Where, in General Relativity, does gravitational potential energy reside, how is its influence felt and how does General Relativity address the "wink in/wink out" aspect of the mechanics of Special Relativity when integrated into the Newtonian approach to gravitation? The author has yet to find a text on General Relatively that even begins to address these issues or to even recognize that they are issues[54].

---

that, so far, has helped those who have learned to recite it avoid having a clear explanation as to how a discrete object at some defined location in the universe may affect the behavior of other objects elsewhere in the universe without using the now disfavored "action at a distance" formulation that Newtonian mechanics and gravitation once employed. To be sure, one of the approaches to the graphical representation of the relationships in Special Relativity between measurements in different "inertial" frames of reference suggests that the measurements in one frame may be determined in another by employing two sets of coordinate axes with one rotated with respect to the other through an angle equal to iv/c. See W.G.V. Rosser, An Introduction to the Theory of Relativity, Butterworth & Co. (Publishers) LTD. (1964) at section 6.3 beginning on page 262.   The author can see where those in the field might expand on this approach so that the rotation idea translates into "curvature". The fact remains, however, that reciting that gravitation is the result of space-time "curvature" is no real answer to anything. It is wordsmithing as a device to evade the obligation to provide an intelligible explanation rather than such an explanation in fact.

54 Further below, the author will demonstrate just what he thinks General Relativity does to incorporate gravitational potential energy within its analysis. In that discussion, the erroneous zero-point convention in Newtonian mechanics will again be significant -- the erroneous convention will be shown to have infected and, ultimately, to invalidate the approach of General Relativity. Simply put, as we will see in detail later, General Relativity takes the wrong portion of the sum of kinetic and potential energy that traditionally was determined to be available when two objects are separated from each other and to have installed this erroneous portion within the object whose field is being constructed. Consider, then, if the author were to drop a bowling ball on to a bicycle pump and thereby compress and pressurize the air in the

The absence of answers to these questions -- or even a realization that these are questions -- coupled with the failure of modern physicists to realize that they use two undeniably inconsistent conventions in assigning zero-point potential energy to electrical and gravitational systems, has left the author with no confidence in the existing state of gravitational physics or its dominant theory, General Relativity. Accordingly, the author has gone back to re-examine the foundations of that theory. The focus of this re-examination, consistent with the discussion above, has been on potential energy, and, specifically, whether the foundation experiments of General Relativity pay proper attention to the inevitable gravitational influence of all forms of potential energy. If the foundation experiments do not always and systematically consider potential energy as a source of gravitation -- and as we will see they do not -- then General Relativity, despite it wide-spread support in the scientific community, must ultimately fail. We will chronicle this failure in considerable detail below.

---

pump's cylinder. The cylinder obviously would now contain previously missing potential energy. Consistent with the discussion above, Special Relativity suggests that the cylinder would be a more potent gravitating body than before the drop and, further, we can fairly say that this increased gravitational potency was obtained by virtue of the force of gravity. The new energy the cylinder now contains, however, is not now "gravitational" potential energy. The only gravitational potential energy now actually within the cylinder is the energy that would be realized if the force of gravity were to compress the cylinder further. One could, of course, generate "gravitational" potential energy by use of the increased pressure that gravity previously allowed us to place within the cylinder but this would be done by allowing the pressure in the cylinder to force the pump handle upward to send the bowling ball back to where it started. At this point, the system of the ball, the pump (and the earth) would again have the gravitational potential energy it started with. If the reader is familiar with the discussions in a variety of texts which detail how General Relativity assigns "mass" to the gravitational potential energy in extended objects, he or she should find the analogy above an appropriate and telling response to those discussions.

Before doing so, however, the author will pursue two additional topics. First, the author will consider the process of construction (and destruction) of a hypothetical galaxy. Then the author will also review a variety of additional Newtonian concepts that appear to have been incorporated into General Relativity but that are inconsistent with the influence of gravitational potential energy as interpreted above. The former diversion is largely in deference to the progression of the author's ideas. The ideas included in this work all grew out of the author's contemplation of problems inherent in galactic construction, and, having developed the discussion at length, the author has been reluctant to discard it. The catalog of problem concepts is a continuation of the discussion above and a prelude to the discussion of perceived flaws in the roots of General Relativity. The author feels that the more clear conceptual errors are illustrated in the existing general understanding of gravitation -- all without descent in to the complex mathematics of General Relativity from which no intuitive understanding has ever escaped -- the greater the likelihood that the re-examination of the roots of General Relativity has validity. Also, the author would caution the reader that, because the author's understanding of the interplay of electromagnetic fields and gravitational "fields" has evolved since the next several sections were written, some of the ideas expressed are obsolete. The author's current ideas regarding the interplay of electromagnetism and gravitation are set forth in a much later section of this work. In the author's view, the following sections retain sufficient value to justify their existence and, further, that removing all of the steps by which the author has reached his opinions risks making it difficult for others to follow the trail (and to identify where errors were made if the ultimate conclusion is that the trail leads nowhere).

6.   Constructing a Simple "Galaxy"[55]

To put the discussion above into practice in the search for "Dark" "Matter" within a large scale structure, consider construction of a

---

55 After the original drafting of this section, it occurred to the author that the discussion which follows assumes that there is no material interaction between the galaxy we are building and all of the other objects and systems in the universe. In truth, then, in the following discussion, we are really building a "universe" and not a "galaxy". In building a galaxy within a universal system, we would have to pay attention both to the potential energy that becomes stored when our objects move away from the galactic center-of-mass and the potential energy realized (or stored) when our objects move toward (or away) from the universal center-of-mass which presumably is located elsewhere and not within this particular system. This may, in part, explain why most galactic systems are not radially symmetrical. If a galactic center is a localized area of enhanced mass that pulls equally on all of the objects that surround this area, the galaxy would still not be not circular in shape because the influence of this local area of attraction is compromised by the attraction of the entire system toward the universal center. There is also the consideration that this area of enhanced mass density may be moving away from a central point (as a result of the event that currently is described as the "Big Bang"). The author cannot help but notice that the large scale systems we encounter in common experience bearing shapes most like those of galaxies are hurricanes, which are areas of low pressure which "attract" the excess fluid in areas of higher pressure. The apparent movement of the area of low pressure on a rotating earth -- and the resulting the Coriolis effect -- generates a system with spiral-like arms and does so because the low pressure area moves with the rotation of the earth and the accelerating currents headed toward the low pressure area are constantly missing their moving target.

hypothetical "galaxy". Begin with a central system of objects in a symmetrical distribution that would capture any object that moves radially outward from the central system at a velocity of less than .9999 percent of the velocity of light[56] and assign the aggregate mass of this system the value "$M_{Central}$". Next, take a symmetrical ring of matter -- a cosmic hula hoop -- that has a rest mass of 1 percent of the mass of the central system and place this ring around the circumference of the central system. Next, by a means that leaves no residue other than kinetic energy[57], instantly accelerate this ring outward radially with a velocity of .999 of the speed of light.

---

56 At some point, assuming that the concept in this work is accurate -- i.e. that potential energy stored in a gravitational "field" itself has "mass" and has its own gravitational "field" even though this "field" is associated with "empty" space -- it should be possible to go back and calculate the number, mass and radius of any set of objects that could make up this central system, knowing that much of the mass of this system will not be "matter" at all but still more stored gravitational energy. Available information, however, suggests that even a single neutron star has an escape velocity of 50 percent of the speed of light. See M. Begelman and M. Rees, Gravity's Fatal Attraction, at page 46, Scientific American Library (Distributed by W.H. Freeman and Company) (1996). It therefore seems likely that even a modest number of neutron stars in close proximity would generate a system with a relativistic escape velocity, even if ultimate analysis suggests that .9999 c is in excess of the escape velocity of any single galaxy and thus is not realistic. Note that, so long as the escape velocity of a system is relativistic, the concepts in this work should have validity, even if the particular example (which builds a large amount of "dark" "matter" quickly) is later show not to represent any real system.

57 Such as use of equal of amounts of matter and anti-matter which, upon annihilation generate the necessary energy. The mass of the combined sum of this matter and antimatter would, of course, be substantial and, in fact, the author has assumed that the combined mass of this matter and anti-matter would be the same as the mass added to the 1 percent ring after the instantaneous acceleration. Thus, in the construction effort, the author is ignoring "exhaust" from the fuel -- the author is assuming that the exterior ring simply pushes on the central system

Immediately after the ring comes up to its initial speed, the inertial mass of the matter in this ring would, from the perspective of an observer located at the center-of-mass of the system, have increased such that the rest mass of the ring (its "traditional" and thus potentially luminous matter) would represent only approximately 4.5 percent of the inertial mass with the remainder (95.5 percent of the total) representing the enhanced mass associated with the movement of this traditional matter at a relativistic speed.[58]

Because this ring of matter is moving at less than the escape velocity of the assumed system, ultimately the ring will come to a stop.[59] When it does, its gravitational and inertial mass will equal its original rest mass of 1 percent of $M_{Central}$. The mass formerly associated with its radial motion is now stored as potential energy in the gravitational "field" of the combined central object-external ring system. The mass of this potential energy is, of course, substantial -- it is 21.36627 times the

---

in a symmetrical way so that the central object's center-of-mass and shape are unchanged. The author is also assuming, certainly without justification, that the central object behaves as a totally rigid sphere that is not deformed by the launch and thus that all of the potential energy in the fuel is taken away by external ring. The system the author is attempting to describe would be like the temporary state in the middle of an elastic collision (of the in-falling ring and a spherical central object in this instance) as described in Tolman, Richard C., Relativity Thermodynamics and Cosmology, Oxford at the Claredon Press (1958), Section 23, page 43-45.

58 The relative percentages have been generated by applying the second of the formula in 5.a., above:

$$m(v) = \frac{m_o}{(1 - v^2/c^2)^{\frac{1}{2}}}$$

59 The author believes that it is still unnecessary at this point to assign a geometry (Euclidian or otherwise) to the space that we are analyzing. An observer in the center-of-mass system of the relevant galaxy will know that the object had to come to a stop because it left the central system with less than escape velocity.

mass of the external ring, and, thus, 21.36627 percent of the mass of the central object. The mass equivalence of this potential energy -- like the mass equivalence of the kinetic energy it replaces -- would appear to fit the criteria for "Dark" "Matter." The only way to sense its presence is to feel its gravitational pull.[60] Further, this "dark" mass represents a material fraction, 21.36627 percent, of the central mass so that, at this point, the ratio between traditional matter and "dark" "matter" in this system is 101 (the starting mass of the central object of 100 percent plus the 1 percent in the external ring) compared to 21.36627 (the mass of the potential energy stored in the object-ring system).[61]

The next step in this process is to give this first ring sufficient speed perpendicular to the line back toward the central object so that the ring will remain in orbit (and can be ignored in our further work). The mass

---

60 Or to allow the mass to fall back to the center and reconvert the potential energy to kinetic energy or to some other form of energy such as radiation.

61 Note that there is a parallel between the assumption that there is mass implied by the lifting of a gravitating object within a gravitational "field" and the recognition that there is additional mass implied by the introduction of additional energy into a simple system held together by electrical forces. Thus, few would argue that the absorption of a photon by a hydrogen atom in its ground energy state would move that atom into an excited state and that the excited atom would have an increased gravitational mass. The mass of the excited system is equal to the masses of the original proton and electron as a consolidated unit (which would be less than the rest mass of the proton plus the rest mass of the electron by the amount of their binding energy) plus the mass of the photon that was absorbed. Because photons of different wave lengths can be absorbed, the two-body hydrogen atom system can therefore have a variety of masses and it is inappropriate simply to add the gravitational influence of the proton to the gravitational influence of the electron to determine the gravitational influence of such an atom, excited or not. In a similar vein, if one used an energy source to separate one gravitating object from another gravitating object, shouldn't the combined mass of the system be greater than the mass of the original objects alone?

equivalence of the kinetic energy given to this ring would be another component of dark matter although, at this point, it is assumed to be negligible because this external ring is likely to be sufficiently distant from the central ring that its orbital velocity is not relativistic.

Our next step is to add another ring of matter inside the first ring that we have launched. Before doing so, it is interesting to consider the nature and distribution of the "dark" "matter" that is now incorporated into the hypothetical galactic structure. The potential energy that provides this inertial mass is now associated with the space through which the first ring has traveled[62] such that each volume of this "empty" space exerts a gravitational pull. If one assumes that mass/energy is conserved in the expansion process, then the mass of each infinitesimal volume of the

---

62 For a similar analysis (actually an analysis of the inverse situation and involving a three dimensional rather than 2 dimensional system) done with respect to a spherical object that is electrically charged, see E. Purcell, Electricity and Magnetism, (McGraw-Hill Book Company 1965) beginning at page 49. Note that the author has ignored the gravitational influence of the components of the external ring on itself because the external ring represents but 1 percent of the total matter in the system. The author thinks, however, that an imaginary experiment in which the only traditional matter is a distribution matching that of the external ring by itself yields some very interesting results. In particular, there should be a gravitational "field" acting within this ring -- produced by the gravitational potential energy stored in the system -- even though there is no traditional "matter" inside this distribution with which one would normally associate such a "field". The same would be true of a three dimensional spherically symmetric mass distribution. In contrast, a spherically symmetrical mass distribution would, under a classical analysis, produce no internal field -- as discussed in D. Klepper and R. Kolinkow, Introduction to Mechanics, W.W. Norton & Company, Inc. (1968), starting under at Note 2.1 at page 101. The construction of a gravitational "field" within a spherically symmetrical mass distribution is discussed in a following section although the discussion is incomplete because the author frankly lacks the sophistication to solve the problem in calculus that needs to be solved, lacks the time to acquire this sophistication and, in any event, believes that the end result is just a more sophisticated approximation rather than a rigorous analysis of gravity.

"field" would be equal to the mass equivalence of the work done in lifting the relevant segment of the external ring through this particular volume. The density just outside the central object is the greatest (because it is there that the deceleration due to the central object is the greatest and thus where more work is required to lift the ring through a given radial distance) and declines with increasing distance. Because of the symmetry of the system, the net pull of this distribution would be toward the system's center with the amount of the pull a function of distance from the central distribution. Accordingly, this mass is smeared outward in a smooth, radially symmetrical distribution of declining density beginning at the central ring and extending to the external ring. This "empty" space, however, is most certainly not devoid of "mass."[63]

Returning to the construction of our "galaxy", next let us take a second symmetrical ring, give it a rest mass of 1 percent of the mass of $M_{Central}$, again place this ring around the circumference of the central mass distribution, and again, by use of an energy source that leaves no residue other than kinetic energy, send this ring moving outward radially with a velocity of .999 of the speed of light.[64]

As before, this matter will decelerate and ultimately stop. The point at which it stops, however, is short of the point at which the first ring

---

63 Contrast this with the suggestion that there is a uniform and constant gravitational "field" within an empty spherical cavity at the origin of a static, spherically symmetric mass-energy distribution or that there is no field within such a distribution. See T. Frankel, Gravitational Curvature (An Introduction to Einstein's Theory), W. H. Freeman and Company (1979) at 53-54.

64 There is obviously some imprecision here. The new ring of matter and the mass of the fuel to propel it could not just appear. They would have to be brought into the system and would assert a perturbing influence (they would pull the first ring in tighter, for example) or they would have to be generated by reducing the mass of the central system and using the mass reduction to "fund" the ring and launch fuel. For the purpose of this example, which is addressing quantitative issues only, we are ignoring these impacts to keep the system as simple as possible.

stopped because of the additional mass (21.36627 percent of the central mass) deposited by the passage of the first ring. At this point, "dark" mass represents 42.73254 percent of the central mass so that the ratio between traditional matter and "dark" "matter" is now 102 to 42.73254.

The next step in this process is to give this second ring sufficient speed perpendicular to the line back to the central object so that the ring will remain in orbit (and so it too can be ignored in our further work). The kinetic energy given to this ring again would be another component of dark matter although, as was the case with the first ring, the mass equivalence of this potential energy is assumed to be negligible.

If we continue with this process and add more and more rings, we can arrive at any relative mix of traditional matter and "dark" "matter".[65] Each mass ring at a distance r from the central distribution is but the "tip" of an "iceberg". Each contributes its own rest mass to the system (by analogy to the portion of the iceberg that is above water) and an additional (and much larger in the case of a ring at a material distance from a mass distribution with a relativistic escape velocity) amount of mass equal to the mass equivalence of the kinetic energy necessary to move this ring outward from the very center of the mass distribution through the gravitational "fields" of:

- all of the rings within this ring's radius; together with
- all of the mass equivalence of the potential energy stored in the gravitational "fields" between those rings; **together with, surprisingly,**

---

65 Except, the author supposes when we add one too many rings such that the escape velocity of the central distribution exceeds the speed of light. Construction of such a system would then stop because further additions are not possible. It appears, then, that the concept of a black hole within a galaxy and gravity as a conservative force are not consistent. This is discussed in a further section below.

♦ a large portion of the potential energy stored within the radius of this ring as a result of the hypothetical possibility of the passage of the outer rings through this space on their way out to their actual locations. The quantity of this mass would have to be inferred from the measurement of the mass of the objects **outside** the radius of this object (consistent with the discussion in footnote 14, above).

The author has not attempted a calculation to see whether such an arrangement would result in a distribution of velocities that is like that of a record on a record player (as the distributions of real galaxies have been described)[66] rather than like the solar system. Intuitively, the author suspects such a result. Indeed, the author takes some comfort in the fact that this curve has more than passing similarities to the curve derived in the ultimate speed experiment for electrons.[67] The slope of the velocity curve approaches but does not exceed a limiting value just as the velocity of a particle with a rest mass can be made to approach but never exceed the speed of light. At some point, there will be a radius beyond which the launch velocity required to reach a further ring position is a sufficiently large fraction of the speed of light that the mass equivalence of the potential energy necessary to move a new ring of matter out to this still greater orbit (considering as well this mass equivalence's impact on the calculated mass-equivalence associated with the greater potential energy necessary to keep rings that are already present at their existing distances) represents a sufficient fraction of the rest mass of the new ring that the added mass associ-

66 See J. Wheeler, A Journal into Gravity and Spacetime, at page 237.

67 See, French, A.P., Special Relativity, W.W. Norton & Company, Inc., Figure 1-3 at page 10 (1968). An explanation of the discrepancy between the curve from the ultimate speed experiment and the velocity/distance profile of a galaxy could be that the density of matter in the outer rings of the galaxy is not constant, but, instead, the rings at the greatest distance grow less dense. The author would also note in passing that the curve has more than a superficial resemblance to the plot of nuclear binding energy to atomic number.

ated with this potential energy necessarily holds the new ring within the system at the appropriate radial velocity.[68] In this regard, remember

---

68 The author suspects that it would be easy to replicate, by computer simulation, the appearance of a real galaxy by trial and error by assuming a system with an escape velocity of say .99 percent of the speed of light, allocating portions of the mass of this system between fuel and rings of traditional matter and then assembling an extended ring structure from the outside toward the inside. If one assumes that each ring has the same total mass and if one picks an initial amount of fuel to use in launching the furthest outside ring, one could then calculate the amount of fuel necessary to place another ring just inside the outer ring and so forth until the fuel is exhausted. Depending on the breakdown between fuel and traditional mass, one could get something that resembles a hula hoop (selected initial fuel too high) and the planet Saturn (selected initial fuel too low). Interestingly enough, the author encountered a system like the hula hoop in a 2005 calendar of images from the Hubble Space Telescope. The calendar is entitled "Images from the Hubble Space Telescope" and was published by Cedco Publishing Company, 100 San Rafael, CA, 94901. The image is the cover photograph and also the photograph associated with the month of July, 2005. It is identified as an image of "Hoag's Object." The most striking features of this system are a large and intense central bright yellow "dot" separated from but surrounded by a blue ring. The brightness of the "dot" falls off with increasing distance from the center until there appears to be no source of light until the blue ring is reached. Picture the object as an archery target with a center of yellow that is most intense in the very middle, then a ring of black and then a ring of blue that includes objects that appear as typical stars. The obvious electromagnetic analogy to this system is the system of components used in a cathode ray tube. In such a tube, a potential difference between an electron source and a grid is created in a vacuum so that electrons generated by the source are then accelerated toward the grid and thus form a coherent beam. Applying this analogy, we would take the system of blue stars as the grid that supplies the potential difference, the electron source would be the matter and radiant energy on the far side of the object and the bright yellow dot would be the analog of the "electron" beam although the system would accelerate electromagnetic radiation as well as

from the end of Section 5 c., above that the addition of one unit of aggregate mass to the point in a system from which the unit would acquire a velocity of .999999999 of the speed of light upon collapsing into the center-of-mass actually represents the addition of 22,360 units of mass to the system.

The similarities between the galactic velocity plot and the velocity plot from the ultimate speed experiment further suggest an alternative derivation of the author's conclusion that the "empty" space within a gravitational system must exert its own gravitational pull. Imagine conducting the ultimate speed experiment in reverse but using a gravitational "field" to produce the required deceleration rather than an electrical field and using identical gravitating bodies rather than electrons as subjects. In the ultimate speed experiment, an electron is accelerated to relativistic speeds and grows in inertial mass as it does so. The enhanced mass associated with this moving electron is drawn from the energy of the electric field that causes the acceleration and, if one ran this experiment in reverse, conservation of energy would require that the enhanced mass associated with the rapidly-moving electron at the starting point of the experiment would be deposited in the electric field that causes the deceleration[69]. When one substitutes a gravitational accelerating system for the electrical system, one should be able to replicate the result -- the enhance mass of an object accelerating in a gravitational "field" should be obtained from the potential energy stored in the gravita-

electrically charged particles. Indeed, the intense yellow dot may simply be red or blue shifted radiation that originates on the far side of the system. The author doubts, then, that there is any traditional object in the center of this system generating the radiation that we perceive as the yellow dot.

69 See, French, A.P., Special Relativity, W.W. Norton & Company, Inc. (1968), at pages 9-10 and J. Jewett, Jr. and R. Serway, Physics for Scientists and Engineers, 6th Ed., Section 25.2 at page 766 (Thompson Brooks/Cole 2004). See also, Tolman, Richard C., Relativity Thermodynamics and Cosmology, Oxford at the Claredon Press (1958), Section 42, page 89-90.

tional "field" and a decelerating object should deposit its enhanced mass in the field as it slows.

Thus, as suggested in a footnote above, it is not surprising that the curve reflecting the velocity distribution of objects orbiting a mass distribution[70] with a relativistic escape velocity looks very much like but is not identical to the curve associating speed with kinetic energy for particles with a constant rest mass. All that must be assumed is that the rest density of the objects orbiting the mass distribution with a relativistic escape velocity declines with increasing distance -- because exterior rings are spread through an ever expanding volume -- to explain the modest differences between the two curves.

In any case, it seems unreasonable to look for the exotic dark matter that people have suggested as explanations of the "missing" "mass"

---

70 See J. Wheeler, A Journal into Gravity and Spacetime, at page 237, bottom graphs.

problem without first fixing the distribution and impact of the dark matter discussed in this work.[71]

---

71 As noted above, General Relativity already attempts to incorporate in its field equations the concept of a gravitational pull associated with potential energy stored in a gravitational "field". The derivation of the measure of this potential energy -- at least as detailed in T. Frankel, Gravitational Curvature (An Introduction to Einstein's Theory), W. H. Freeman and Company (1979) at 29-31 -- appears far more complex (and subject to miscalculation) than the rather simple derivation suggested above. The above approach measures gravitational potential energy simply by assuming that an object at a distance r away from a large mass (with an internal structure that can be ignored) got there by being launched from the surface of the large mass and measuring the mass equivalence of the kinetic energy necessary for the object to reach the distance r. Because the system is viewed consistently from the point of view of an observer at rest with respect to the center-of-mass of the large mass/object system and, thus in an area free of gravitational "fields", there should be no relativistic complications to the potential energy measure suggested by this methodology. A critic, of course, might question the propriety of such a "universal" frame of reference given the tenets of General Relativity. The author's response is that this same frame of reference is implicit in the formulation of the "Dark" "Matter" problem. It is therefore impossible to simultaneously see such a problem, frame it as it has been popularly framed and to question the use of a single, universal center-of-mass frame as the one in which the problem can be solved. Further, at least one commentator on the concepts of Special Relativity has proven the existence and universal nature of a unique center of momentum and center-of-mass frame of reference. See Rindler, W. Essential Relativity, at Section 5.7 (Springer-Verlag 1977), where the author appears to establish: (1) that there is a unique center of momentum frame of reference for any system of objects; (2) that this center of momentum frame is also the center-of-mass frame: (3) that the mass in the center-of-mass or center of momentum frame corresponds to the "rest" mass of the system if its composite nature were not recognized; and (4) that observers on all objects in the system would agree on the "mass" and "velocity" of the center of momentum frame.

7.   Concept Check, "Galactic Collapse"

As a check on the galactic construction process, consider the collapse of the first shell into the central distribution before any further shells are added as viewed in the frame of reference of the system center-of-mass. As each segment of the distribution falls toward the center, it picks up kinetic energy and its inertial mass increases according to the formula obtained in Section 5.a., above. Because the sum total of the mass and energy of the system is assumed to be conserved and because the inertial mass of one form of energy must equal the inertial mass of any other, the enhanced mass associated with the now falling ring must come from somewhere. The place where it must come from to avoid instantaneous action[72] at a distance if for no other reason

---

72 Avoiding instantaneous "action at a distance" was assumed necessary in the development of General Relativity and the author finds the arguments in support of this idea generally convincing. Nevertheless, it seems to the author that there are problems inherent in any system that attempts to reconcile the following three concepts: (1) a fixed velocity for the transmission of the gravitational force that is equal to or less than the speed of light; (2) gravitational mass and thus presumed gravitational interaction between two photons moving in completely opposite directions as would be the case in an electron-positron annihilation; and (3) conservation of mass/energy. A finite speed of interaction between the photons limited at the speed of light would suggest that neither could pull on the other as they moved apart. No signal from one could ever reach the other. On the other hand, if the photons are absorbed

is from mass associated with the gravitational potential energy formerly located in the region through which the fragment of the shell has just fallen. In other words, since there certainly is inertial (and therefore gravitational) mass associated with kinetic energy (based on the well tested formula above), conservation of mass/energy requires that an equal amount of positive gravitational mass must be associated with gravitational potential energy. The amount of this mass in a given volume of space must equal the mass lost while rising through that space against the gravitational "field" operating over the space or the mass gained while falling back through that space.

Ultimately, upon collapse of the entire system[73], there will be a collision of the various components of the ring at the system's center-of-

---

after they have moved away from each other and the objects that have absorbed them are allowed to collapse into a single object, gravity would do work on the collapse that was not done in the process of separation and thus the final assembly would be more massive than the constituent parts. The author will come back to this idea later.

73 We have spoken of this process as the collapse of this system and our discussion mimics what appear to be contemporary views of the "collapse" of large scale structures. This notwithstanding, it must be realized that all of these discussions ignore a significant dimension to the new approach to gravitation urged in this work. Specifically, every discrete component of a collapsing system -- every proton, neutron, electron and all other components of the chemical elements of which that system is made -- is not only "falling" inward but also moving away from all of the other mass in the universe. Since the accelerating components are growing in mass as viewed from an appropriate reference point (such as the center-of-mass of the collapsing system), their gravitational interaction with all of the other elements of the universe as measured in that reference frame will grow and grow until the interaction with the remainder of the universe prevents further collapse. The collapse to a dimension-less point that is a current feature of gravitational analysis -- the existence of "black" "holes" -- ultimately cannot survive scrutiny. A formal analysis and refutation of the existence of such objects is presented further below.

mass. If the collision is an inelastic one, the total mass of the collapsed in-falling ring will be greater than the rest mass of the components of the ring by the mass-enhancement calculated from the formula in Section 5 a., above.[74] In such an inelastic collision, this additional mass will remain near the center-of-mass of the system -- probably as incoherent kinetic energy (i.e. "heat") but perhaps could be in some other form such as the reversal of an exothermic chemical reaction to create chemical potential energy. If the collision is elastic, there will be an instant at which the entire system is at rest, and, at this moment, the object at the center-of-mass of the system has an identical mass to the object that results from an inelastic collision.[75] The only difference would be that the former kinetic energy of the in-falling components of the ring must be stored as coherent rebound potential energy. Regardless, however, it cannot be possible to model a radially symmetrical object or a spheri- cally symmetrical object with a simple object of identical mass located at the larger's object's center point. The mass equivalence of the po- tential energy required to assemble the larger object forbids this ap- proach, and, of course, the greater the radius of the ring or sphere, the greater the mass of the system of which they are a part so long as the amount of traditional matter in the ring remains the same.

To model the process described above, one can consider a roll of toilet paper and associate the central cardboard core with "traditional" matter and the toilet paper itself with the additional "dark" inertial mass associated with the core's movement. A huge range of ratios of toilet paper to core is possible, depending on how much of the roll has been used. Similarly, a huge range of traditional matter to inertial mass is possible depending on an object's speed, and, in fact, the range of rest mass to inertial mass is much greater than in the author's surrogate

---

74 See W.G.V. Rosser, An Introduction to the Theory of Relativity, Butterworth & Co. (Publishers) LTD. (1964) at section 5.7 beginning on page 217

75 Tolman, Richard C., Relativity Thermodynamics and Cosmology, Oxford at the Claredon Press (1958), Section 23, page 43-45.

-- the amount of inertial mass associated with an object can approach infinity because the denominator in the relevant equation tends toward 0.00 with increasing velocity.

If one places a series of new toilet paper rolls in a circle, pins the loose ends together at the central point and rolls them away from the central point, each will leave a trail of paper (dark mass) as each loses paper (i.e. decelerates in the corresponding gravitational "field").[76] At some point, the "dark" mass associated with the network of paper starting at the origin dominates over the cores and the remaining unrolled paper associated with them.

If one starts with rolls that are almost empty, one would get a system like our solar system in which the cardboard cores are the dominant source of mass. If one starts with full rolls (to model a large galactic system), it is the toilet paper and not the cardboard that ultimately exerts the dominant gravitational pull.

---

76 Of course, the toilet paper would unroll at a constant rate while the energy loss of an object rising in a gravitational "field" would be greatest at the start and would fall off with distance. Nevertheless, the author has found this example helpful and wanted to include it in the hope that it will help the reader as well.

8. Changes Required to Long-Established Classical Concepts

a. A More Complete Catalog of Problem Concepts in Current Theory

If gravitational mass is attributable to the potential energy stored in a gravitational "field" as the author has described above, many of the most long-established concepts of Newtonian Gravitation cannot be true. Some flawed concepts were addressed above and others are detailed below. The cataloging of these faults is necessary in the author's view in part so that the errors are not further perpetuated -- it is dangerous to incorporate such conceptual errors in the core curriculum of all that are to be trained in the discipline.

The author would hasten to add that he finds the many conceptual faults in the present core curriculum as compelling confirmation that no current physicist truly understands the approach to gravitation that they attempt to apply -- no person who understood the impact of gravitational potential energy on gravitating systems would tolerate the continued teaching of the concepts referenced below other than with prominent warnings that these concepts only apply in the limiting case of weak gravitational "fields" operating over modest distances.

Indeed, when the unknowing perpetuation of these errors is coupled with the unrestrained reverence that all working in this field seem to have for relativity and its generalization, it becomes impossible to have a principled discussion of gravitation. At present, in this discipline, there are a select few who purport to have mastered the intricacies of Special Relativity and an even more select few who purport to have mastered the still greater intricacies of General Relativity. These worthies are unanimous both as to the brilliance of the relativity theories and, perhaps even more so, their own brilliance for having mastered the complex and beautiful "insights" of these works of human ingenuity[77]. These doyens endeavor, even now, to weave ever more obtuse extensions of the elaborate fabric of space-time in a determined effort to conceal its glaring failure. They apparently would rather keep the theory and ignore the very real universe that we can perceive. In this effort, they favor all manner of strangeness and will not even entertain for a moment the possibility that the beautiful fabric they have mastered does not exist. Indeed, the more strange the predictions of their theory, the more delighted physicists appear and the more confident they seem to be their theory is "correct."[78] The understudies in this discipline, of

---

77 The author, by these comments, does not mean to suggest that Special Relativity is anything other than a tremendous work of human ingenuity and insight. The author further does not question the propriety of Albert Einstein's determined attempt to generalize it. It is clear, however, that, as a result of the substantial development of Special Relativity and of physics in general since 1905, Einstein's attempt to generalize the relativity theory has failed. The author's dissatisfaction, then, is directed at those who refuse to even entertain the notion of its failure and who suppress all independent thought. Aggressive suppression of open discussion is nearly always a sin but it is an ever more glaring transgression in this discipline. It is in the history of physics that we see perhaps the most outstanding examples of the fruitfulness of independent thinking -- of Galileo, Copernicus, Brahe, Kepler -- over unthinking allegiance to dogma.

78 With the decline of established religion, those in the sciences appear to have replaced the ancient presumption that all of creation is a rational construct by an omniscient architect and builder with a presumption that nature is obtuse, irrational and random. This replacement has little to recommend it. From the dawn of history, science has been a search for

course, simply take the dominant theory of gravitation on faith rather than based on a thorough understanding of it and who can blame them. Who working in this discipline -- and hoping to obtain employment or grants for independent study -- would dare step forward to say that he or she cannot "see" the marvelous fabric that is space-time? The author, having read extensively in the literature of Special and General Relativity, can only sympathize with these understudies. This literature is now littered with needless mathematical generalities, complex and abbreviated mathematical presentations and jargon devoid of meaning or any value in developing an intuition regarding phenomena.[79]

---

order, rationality and simplicity. This search has borne such substantial fruit that its fruitfulness alone provides justification for the belief that the universe is organized and governed by reasonable rules of limited complexity. Of course, every presumption may be overcome by the facts in a particular case and we should not allow ancient (or modern) prejudices to overrule reason.

79 Does the attribution of gravity to space-time "curvature" really convey any meaning? At all? Do discussions of "gauge" transformations really provide enlightenment or rather obscure simple concepts -- concepts that ultimately the author thinks will be found deficient -- under jargon that, deliberately or not, prevents non-insiders from easily understanding basic ideas? Indeed, is there anyone who can really follow a tensor presentation using curvilinear coordinates and the Einstein Summation Convention into any meaningful conclusions? The author has his doubts. If the present work has value, then one can only conclude that no one in the past century or so has been able to use existing formulations of gravitation and many aspects of the formulations of electricity and magnetism to reach meaningful conclusions about anything touching reality. In this regard, it probably is not a coincidence that most of the useful books the author has read are old. The writers of these books apparently assumed that the concepts they were introducing would be difficult to grasp for even the best students, and, likely for this reason, progressed slowly under the assumption that the reader had the right to demand simplicity, transparency and clarity as well as some effort to avoid pure abstraction when dealing with such difficult material. In contrast, the more recent books the author has attempted to read -- usually with limited success -- have been written under the assumption

As a counterpoint, the author suggests a reading of Hans Christian Anderson's "The Emperor's New Clothes." To the author, the fabric of the emperor's clothes and fabric of "space-time" are the same -- some amalgam of intellect, pride and hubris. Both fabrics acquire their dubious substance because members of groups with special status and expertise fear the exposure of their lack of understanding more than the abetting of the absurd. Indeed, a good case can be made that the most dangerous people that have ever been and ever can be are the "experts" that are so common in modern times[80]. It takes an "expert" confident in his craft to do the absurd with unshakeable confidence. The only defense against the tyranny of a such an educated elite is a constant demand that they explain, in simple terms, the results of their labors. The elite, of course, is likely to protest that the common stripe of man is just not intelligent or sophisticated enough to breath the rarefied air in which they work. This is nonsense. None of us is really so clever and no one's credentials are so impeccable that his or her analysis must be accepted without question.

Indeed, physics today is so firmly fixated on credentials and not ideas that it is blind to the intense irony evident in its history. The physics and mathematical credentials of the architect of General Relativity, Albert Einstein, are clearly deficient by modern standards such that, were he to step forward today with new insights, it is doubtful that his voice would be heard. The need to question the authority of "experts" is a lesson that mankind has seemed to learn and unlearn many times. Who can doubt the tragic confidence of the marine architect of the

---

that the writer has the right to demand almost infinite time, attention and intellect from his or her readers. These writers adroitly avoid any challenge to their expertise. If no one can discern what they have said with sufficient clarity, then none can criticize it.

80 Remember that Karl Marx, the architect of one of the most destructive approaches to the organization of industrial and other productive systems, was a writer on economics and still draws substantial support in academic circles even today.

Titanic in designating his creation as "unsinkable" and, thus, the danger that lurks when we lose sight of our own limitations. The author hopes that the lesson of what should be the rapid fall of "General Relativity" from established doctrine to disrepute will allow us to relearn and re-member the fallibility of experts -- if, perhaps, only for a while.

Continuing on, then, with a cataloging of the most dangerous of the problem concepts of Classical Gravitational that remain part of the what physicists learn and believe, let the author suggest that the following ideas are rendered obsolete by the mechanics of Special Relativity:

- That the gravitational "field" inside a spherically or radi-ally symmetrical distribution of mass is necessarily zero. Instead, as discussed in detail in the following subsection, there is a "field" resulting from the potential energy associ-ated with the movement of the components of the distribu-tion out from the center-of-mass of the distribution to their current locations. Indeed, this "shadow" "field" may be very substantially more intense than the "field" associated with the visible components of the system -- the shadow "field" depends on the amount of fuel consumed to produce a giv-en distribution of mass. Accordingly, even the tiniest trace of dust, if ejected outward as a result of an explosion of fuel of sufficient mass, can create a hypermassive system. Further analysis of this situation is contained in Subsection 8.b., below.

- That it is possible to define a gravitational potential "field" that can be used to calculate the force on any object, regard-less of its mass, inserted at any specified point. Instead, the insertion of a mass at a point a given distance from a ref-erence object necessarily implies the insertion of sufficient (and non-linearly variable) mass in the form of potential en-ergy into the space between the reference object and the in-serted mass so that mass/energy can be conserved as the inserted object falls into the reference object. The perturbing influence of this additional mass cannot be ignored if the object inserted is massive or, regardless of mass, if the object is inserted toward the edge of a system with an extreme relativistic escape velocity. Further analysis of this situation is contained in the author's discussion of the problems with General Relativity below since this erroneous proposition is essentially the "Principal of Equivalence" that is the soul of General Relativity. Note that the author's challenge to this proposition necessarily means that the concept of a gravitational "field" as it currently is understood is no longer defensible. For that reason, the references in this work to a gravitational "field" use the word in quotation marks. Only weak gravitational "fields" behave like true fields.

◆ That it is possible to imagine a "test" mass that can be moved from point to point within a gravitational "field" to establish the strength of a gravitational "field". Instead, even the most inconsequential test object, if inserted a material distance away from a system with a relativistic escape velocity, implies insertion of additional mass that tends toward an infinite amount as the reference object's escape velocity approaches more closely to the speed of light. Moreover, a test mass cannot be "moved" within a gravitating system except by the expenditure of a fuel for outward motions or an accumulation of the mass associated with kinetic energy for inward motions and, thus, the movement of a "test" mass invariably results in a disturbance of the system that is non-linear and which tends to infinity. Accordingly, even an attempt to use "identical" electrons as test masses will fail because an electron at a point on the fringe of a hyper-massive system represents a substantially greater investment in system energy than an electron near the center-of-mass of the system. If the fringe electron were allowed to fall to the point at which its test companion is located, the only similarities would be their conserved charge. The result of a collision between the rapidly moving in-falling fringe electron and an electron deep within a gravitating system would be a series of complex particles just as the acceleration of an electron in a particle accelerator (using an electric "field" to increase the speed of the electron and magnetic fields to contain it) and its collision with a target yields something that behaves quite differently than an electron.

- That the gravitational potential at a point associated with a distribution of masses can be determined by the superposition principal -- the addition of the gravitational potentials associated with each discrete mass. Instead, each additional mass implies exponential growth in the "field" strength. This is the case because the addition of a new gravitating object to a gravitating system at any point necessarily requires:

  - if the object is added to the system at the center-of-mass, the addition of a further amount of mass required to move all of the existing components of the system out to their existing locations against the pull of the objects that were already present and the still greater pull of this object and against the potential energy implied by the insertion of this new object and the greater potential energy associated with the radial distances at which existing objects are located.

  - if the object is added to the system at other than the center-of-mass, the addition of an additional amount of mass in the form of the potential energy required to move this object out to the relevant point against the gravitational "fields" of all other objects and the potential energy associated with the distribution of those objects (which is further increased by the additional potential energy associated with the existing positions of each of the existing components).

- That, very surprisingly, if two test masses of differing magnitude are each, in turn, placed at a defined "distance" from a gravitating body and they are pulled on by forces in a given ratio, then the inertial masses of these test "objects" have this ratio. Instead, the ratios of the gravitational pull to inertial mass will vary depending on the mass of the reference gravitating body with the divergence becoming measurable if the reference object has a relativistic escape velocity.[81] Thus, if the ratio of the inertial masses of two bodies is 1:2 and the ratio of gravitational mass of two bodies is 1:2 in the "field" of an object with a non-relativistic escape velocity (like the sun), then the ratio of inertial masses will remain 1:2 but the ratio of gravitational masses will inevitably be larger (at least 1:2, of course, but tending toward 1:∞) when

---

81 The author suspects the immediate reaction to this assertion is that it cannot be true because it is contrary to incredibly precise experiments performed by Etvois and Dicke and other experiments that have been performed through the years. In all these experiments, however, the differing masses are always located very near to each other and thus the potential energy implied by each mass' location within a "field" acts on both masses all of the time. The best analogy the author can come up with to illustrate this point is as follows: consider two "icebergs" frozen out from a single fluid (say water) and with identical profiles above the surface of two differing fluids (say Carbon Tetrachloride and Hexane or metallic Mercury) in which they float. The natural supposition is that these icebergs are identical. If the fluids in which they float have differing densities, this supposition would be wrong. The object floating in the denser fluid must be smaller. It is only when the fluids in which our "icebergs" float have identical densities that we can say that two objects with the same above-surface profile have the same below surface profile as well. Further we can say that an object that has twice the above-surface volume than a second object has twice the mass of that second object only when the densities of the two fluids are the same. When these objects are floating in fluids of different densities we have to take the differing densities into account in determining the mass of each object.

in the fields of objects with increasingly relativistic escape velocities.

♦ That kinematic computations can be made without considering the escape velocities of objects (i.e., the same formulas cannot be used to trace the trajectories for the collision of two hydrogen molecules and the collision of two neutron stars).

b.   Correcting the Problems with Established Concepts -- the Gravitational "Field" Associated with an Outwardly Expanding Sphere Reconsidered

One of the oldest and most widely known expositions of currently accepted views on gravitation -- and thus one of the most significant syntheses of established concepts -- is the Newtonian demonstration of the gravitational influence of a spherical shell of material. Newton's analysis of this situation is repeated in one form or another in every book on mechanics the author has read[82] and is even reproduced in a very well known text on Calculus[83]. To those that have seen the analysis, its ingenuity is remarkable and its outcome is almost certainly remembered -- a spherical shell of material with a total mass of x behaves as a point mass of quantity, x, to observers outside the shell. To those inside the shell, the exterior shell might as well not exist. According to Newton, there is no gravitational influence to be felt inside such a system. The ingenuity of this analysis and its wide acceptance notwithstanding, however, the author believes that it also must be a casualty of a proper understanding of gravitating systems.

---

82 See, D. Klepper and R. Kolinkow, Introduction to Mechanics, W.W. Norton & Company, Inc. (1968), under Note 2.1 at page 101.

83 See, M. Kline, Calculus An Intuitive and Physical Approach, 2nd Ed., Dover Publications, Inc. (1998).

To see why and further expand on the discussion above regarding the complications inherent in the mass equivalence of gravitational potential energy, imagine a uniformly thin symmetrical distribution of matter with density $\sigma(R_{final})$ located on the surface of a sphere with R_{final} as its radius.[84] The mass distribution suggested would, in a classical sense and as indicated above, represent the gravitational equivalent of a "Faraday" cage when dealing with electrical fields and we would expect, based on classical concepts, that the gravitational "field" inside this distribution would be zero just as the electric field inside a "Faraday" cage is nonexistent. If one considers, however, that the potential energy stored in this mass arrangement also has mass and that this mass also has an associated gravitational "field", a net gravitational "field" must exist inside the distribution. The properties of the "field" associated with this distribution will be quite distinct from the properties of the "field" associated with the traditional matter that spawned it, however.

To see that this is the case, assume that the existing distribution was created by placing, at the center of the distribution, the amount of traditional matter currently observed in the distribution (M_T) along with an additional amount of mass in the form of a fuel that burns completely (M_F) such that this fuel is just adequate to lift M_T out to the radius R_{final}. To those outside such a system, such an assembly would behave as a point mass with the total mass $M_T + M_F$.

We now ignite the fuel and the distribution will expand outward from the center much as a balloon that is being inflated expands. The mass of the entire system prior to ignition and the mass of the expanding sphere immediately after ignition would be $M_T + M_F$ from the

---

84 Hereafter, the author will assume that the space in which we are operating is Euclidian and that our time coordinate is based on a frame of reference with its origin at the center-of-mass of our distribution. The clock at this location would be free of gravitational influences except from outside our system and would be "stationary" throughout the entire expansion of our sphere. The author will justify the selection of this reference system in the discussion which follows regarding problems with General Relativity.

perspective of an observer at rest at the center of the distribution (who, incidentally, would not be in a gravitational "field" of any kind other than a field generated by objects outside the sphere). Immediately after the fuel is burned, this mass would entirely be associated with the moving traditional matter so that, by the first of the formulas above, the mass attributable to the expanding distribution would be

$$M_T + M_F = \frac{M_T}{(1 - v^2/c^2)^{1/2}}$$

Using an approach similar to that taken in the discussion of the force on a uniform surface charge as set forth in E. Purcell, <u>Electricity and Magnetism</u>, (McGraw-Hill Book Company 1965) beginning at page 49 -- although recognizing that our situation is the inverse of this example since it takes energy to assemble an electrical sphere and it takes energy to separate a gravitating sphere -- we would introduce a mass density on the surface of the sphere so that the total mass of the distribution at the start is equal to $4\pi r^2 \sigma(r_i)$ and the gravitational "field" strength just outside the sphere would be $4\pi\sigma(r_i)$. The total "field" strength inside the distribution would (in a classical sense anyway and to a reasonable approximation in our analysis *if* r_i is incredibly small)[85] be 0, so the effective gravitational "field" strength of the distribution on itself would initially (and in a first approximation) be the result of ½ of the total mass or $2\pi\sigma(r_i)$ (because the force on the inside of the distribution attributable to the mass on the outside of the distribution is initially zero given the character of classical gravity as a central force that declines as the square of the distance).[86]

---

85 As suggested above, the assumption that there is no gravitational "field" inside a uniform sphere likely fails when one considers that gravitational potential energy has its own mass such that the potential created as the exterior of the sphere expands likely does ultimately exert a force on the interior.

86 <u>See again</u>, D. Klepper and R. Kolinkow, <u>Introduction to Mechanics</u>, W.W. Norton & Company, Inc. (1968), under Note 2.1 at page 101.

Again using the Purcell approach, as the distribution moves outward to radius, r + dr, each incremental patch of area must rise the distance dr against the force on the patch. The force on this patch using the Purcell model would be the gravitational "field" strength (which is $2\pi\sigma$) times the mass of the patch (σ dA) or $2\pi\sigma^2$ dA. The problem with following the Purcell analysis too closely at this point is that the mass in the sphere declines in absolute terms as it rises rather than merely becoming distributed over a larger sphere -- because the relativistic velocity enhancement declines as the sphere decelerates during its expansion -- and, thus, the overall mass of the sphere is not a constant as in the Purcell analysis. On the other hand, the mass that has been lost by the sphere has been gained by the space through which the sphere has risen which now has a mass distribution that behaves as if the lost mass were a compact point source located at the center-of-mass of the system. The force on the expanding sphere is thus the sum of the ever growing force exerted by the compact point source (which does not exist in the Purcell analysis) and the ever declining force of the sphere on itself (which is what is addressed in the Purcell analysis).

The force per unit area exerted by the expanding sphere on itself is $2\pi\sigma_{(r)}^2$dA. The force per unit area exerted by the mass that has already been deposited within the sphere is equal to the total mass of the initial system less the mass that now remains in the expanding sphere times the mass of the patch or $[(M_T + M_F) - 4\pi r^2\sigma_{(r)}](\sigma_{(r)}dA)/r^2$.

Both of these forces act through a distance dr so that work in the following amount must be done against the first of these forces in causing the entire sphere to expand to a greater radius:

$$dW_{\text{Sphere on itself}} = (4\pi r^2)(2\pi\sigma_{(r)}^2)\ dr = 8\pi^2\sigma_{(r)}^2 r^2 dr$$

To overcome the second of these forces, additional work in the following amount must be done during the same portion of the expansion:

$$dW_{\text{Compact point source}} = \frac{[(M_T + M_F) - 4\pi r^2\sigma_{(r)}]\,(4\pi r^2)\,dr}{r^2}$$

Accordingly,

$$dW_{\text{Total}} = [8\pi^2\sigma_{(r)}{}^2 r^2 + (M_T + M_F)4\pi - 16\pi^2 r^2\sigma_{(r)}]dr$$

As a first step in solving this problem, the author suspects the following substitution would be both possible and useful:

$$4\pi r^2\sigma_{(r)} = \frac{M_T}{[1-(dr/dt)^2/c^2]^{\frac{1}{2}}}$$

or

$$\sigma_{(r)} = \frac{M_T}{4\pi^2 r^2[1-(dr/dt)^2/c^2]^{\frac{1}{2}}}$$

This substitution should be possible because the total mass of the expanding sphere must always be the original traditional mass as adjusted to reflect its ever declining relativistic velocity enhancement.

Accordingly, the earlier expression could be rewritten:

$$dW_{\text{Total}} = \frac{M_T{}^2 dr}{8\pi^2 r^2[1-(dr/dt)^2/c^2]} + (M_T + M_F)4\pi dr - \frac{4\,M_T\,r^2 dr}{[1-(dr/dt)^2/c^2]^{\frac{1}{2}}}$$

Even though the author has not solved this mathematical expression for the gravitational behavior of an expanding sphere, the author feels that it is nevertheless possible to describe some surprising quantitative features of that "field". Thus, for example, assume that there is an observer (with trivial mass) at rest at the center point of this distribution. This observer is subject to no gravitational "fields" because the original expanding sphere of mass pulls the observer equally in all directions and so too does the mass stored in the gravitational potential "field" spawned by it (like an observer in the gravitational equivalent of a Faraday cage except, in this example, there is the original cage and its "field" derived shadow). If this observer moves in any direction, however, he would experience a gravitational force resulting from the mass equivalence of the potential energy stored in the gravitational "field" left as the original sphere of mass moved away from the origin. In other words, the observer could move out of one cage but not the other. This "field" would always pull the observer toward his original location even though there is no "matter" there to generate the pull. Accordingly, if this central observer were displaced from equilibrium, he or she would move as if falling through a hole in a solid mass.[87]

Also very interestingly, in such a system, there would be a continuing transfer of the virtual mass deposited by the deceleration of the outer surface of the sphere -- whose net gravitational pull on interior surface of the sphere would originally be non-existent given the behavior of forces proportional to $1/r^2$ -- from the area outside the sphere to the area within the sphere. Thus, with the passage of time, the effective mass pulling on the interior portion of an outwardly expanding sphere would increase even though the total mass of the system and the rule that is used to compute forces at different distances would not change.[88] The author thinks that this is interesting because of its obvious relevance to the "Dark" "Energy" problem also discussed in an the opening section above. If the pull of gravity were assumed to be static as is the case presently, then the growth of the strength of gravity as

---

87 See J. Wheeler, A Journal into Gravity and Spacetime, Beginning at page 55.

88 To visualize this fact, consider taking two rolls of toilet paper, put one roll on the ground and pin its loose end in place, roll this roll outward just a bit and put a second roll of toilet paper down where the first one used to be and also pin its end in place. If one rolls both rolls out in the same direction, an ever growing portion of the paper that was outside the inner roll is now found inside the inner roll.

the author has outlined would be misinterpreted. An observer who suddenly awakened to the system the author has described might find objects further apart, moving faster and decelerating more slowly than would be assumed based on a current calculation of object velocities and object decelerations due to gravity. This observer might then assume that there must be a force that is counteracting gravity, when, if fact, the rule for determining the force of gravity is constant but the effective mass pulling on some objects is, in fact, increasing rather than decreasing.

Also interestingly enough, if this central observer were to send a signal of a given wavelength outward, an observer (without mass) at rest at a distance "r_1" from the central point would perceive a red-shift in the signal because the signal rises through a gravitational "field" from the moment it is sent. This second observer would therefore assume that the central observer was moving away from him. Further if the second observer were attempting to measure the distance from the central observer to himself by multiplying the inferred current velocity of the central observer by some perceived age of the universe, the second observer would err because there is absolutely no relationship between the red-shift and the central observer's velocity. The red-shift is entirely gravitational and not due to relative motion. The result would, in many cases, overstate the distance.

If there were still a third observer at rest at a distance r_2 that is greater than r_1, there would be a still stronger gravitational "field" between this observer and the central observer because the mass associated with the potential gravitational energy in the area between the sphere with radius r_1 and the sphere where the signal photon happens to be at a given time acts as if it were a massive object located where the central observer is located. This third observer would see a greater red-shift than the second observer (because still more of the energy of the signal photons would now have been lost working against the gravitational "field" of the mass equivalence of the potential energy deposited by the passage of the original sphere). This observer would thus assert that the central observer is moving away from him faster than the second observer even though all three observers are, in fact, at rest. A distance measurement by this observer that again is based on multiplying an observed current velocity by some perceived age of the universe will again be in error.

c.   Correcting the Problems with Established Concepts -- Further Demonstrating the Need to Discard the Principal of Equivalence

As also indicated above, the concept most central to the theory of General Relativity -- the "principle of equivalence" -- should be a casualty of the proper understanding of "Dark" "Matter." Because this assertion is sure to be controversial, the following detailed example is provided as additional support.

The appropriate gravitating system to assist us in reaching this conclusion is typically introduced as an exercise in many first year texts on mechanics and gravitation. This system consists of a radially-symmetrical, planar system of mass -- a "ring," like a hula hoop or bagel -- located a fixed distance -- we will call it the distance "a" -- from a central point[89]. The typical discussion considers the behavior of an object located above or below the plane of the ring and on the axis perpendicular to the ring and passing through the central point.[90] Such an object accelerates toward the center point and, if the distance from

---

89 As suggested above, the author has seen a picture of what appears to be just such a system reproduced in a 2005 calendar entitled "Images from the Hubble Space Telescope" published by Cedco Publishing Company, 100 San Rafael, CA, 94901. Refer to footnote 68, above for a further description.

90 See J. Jewett, Jr. and R. Serway, Physics for Scientists and Engineers, 6th Ed., at page 417, Exercise 53 (Thompson Brooks/Cole 2004)

the center point is not too great, executes simple harmonic motion -- it is accelerated toward the center by a force that varies linearly with the distance from the center, it passes through the center with a speed dependent on the mass in the ring and decelerates until it reaches a point on the other side of the ring that is as far from the center as when it started. Note also that, even if the object is displaced from the center point by too great a distance for the motion to be "simple" harmonic motion, it still passes back and forth through the center until it reaches a maximum distance on either side and this maximum distance is, traditionally, simply a function of the mass in the ring[91].

This system has been selected because the mass in the ring can be varied within substantial limits so that we can pick a system that is so massive that it would accelerate or decelerate familiar objects moving along the system's axis of symmetry to or from relativistic speeds, yet, at the same time, we do not need to worry about a "collision" between our test object and the system or about the return of our test object to the system "surface". Accordingly, we can use the center-of-mass analysis of a collision of two objects having relativistic speeds that we have referenced several times before[92] and yet not consider the complications inherent in the substructure of those objects. Given sufficient mass in the ring structure, we can accelerate even simple electrons, which are generally considered to be "fundamental" particles without

---

91 See E. Purcell, Electricity and Magnetism, Section 2.6, pages 43-48 (McGraw-Hill Book Company 1965) for an analysis of the same problem solved with respect to an electrical system. The solution to the electrical problem, of course, mirrors the solution to the classical formulation of the gravitational problem. Although this work calls into question the future use of such parallel solutions, it is appropriate to consider them here.

92 See, French, A.P., Special Relativity, W.W. Norton & Company, Inc. (1968), pages 172 to 175; and See W.G.V. Rosser, An Introduction to the Theory of Relativity, Butterworth & Co. (Publishers) LTD. (1964) at section 5.6 beginning on page 207 and 5.7 beginning on page 217.

internal structure, to speeds so fast that their masses in the center-of-mass frame of our system can be as great as we please.

In our analysis, we will impose a rule that has a common sense appeal but that has yet to rigorously be enforced in any gravitational analysis the author has seen to date -- we will require that our model be a complete system and that no mass or energy pass its boundaries throughout each of our hypothetical experiments.

With this system in mind, let us first attempt to mark distances along the central axis and let us start this process by placing two identical "test" masses -- we need two so that we can conserve energy and momentum in our work -- adjacent to each other at the system center. Our plan is to launch these two test masses out along the central axis at less than the escape velocity of the system and to identify distances from the system center using the mathematics familiar to ancient surveyors and with input position measurements generated by transmission and receipt of returned, reflected pulses of electromagnetic radiation initiated and received by observers on the ring. Our observers will send out radar-like signals that will be reflected back by our test masses as each moves along the axis and we will measure the angle between the plane of the ring and the antenna or telescope that receives the return signal from each "test" mass[93]. If we know this angle, we can use this information plus the distance between the center of the system and our ring (which we identified as "a", above) plus the fact that the central angle is a right angle to determine where our "test" mass is at an appropriate time[94].

---

93 For the moment, we will ignore the fact that the signaling system we have chosen would perturb the system we are attempting to measure. Ultimately, the fact that the measurement system would cause such perturbations just further confirms the final conclusion of this section.

94 See W.G.V. Rosser, An Introduction to the Theory of Relativity, Butterworth & Co. (Publishers) LTD. (1964) at section 3.6 beginning on page 103 with special emphasis on pages 107 to 108

In the discussion that follows, we will not attempt to describe distances in the ususal way -- by reciting the distance reached as equal to multiples of a "rigid" rod incorporating a distance standard. Instead, we will match distances to initial velocities. Thus, we will provide a certain velocity -- a certain amount of kinetic energy -- to our test object and will mark the point of maximum separation between the test mass and the system center. Consistent with this method, then, we will say that the distance "b" from the center point is the maximum distance between the system center and our test object when that test object left the center point with a velocity of .95 percent of the speed of light; the distance "c" is the maximum distance between the system center and our test object when that test object left the center point with a velocity of .995 percent of the speed of light, etc.

To address problems inherent in the fact that our system will be quite large and thus that the finite travel time of our signals will complicate our analysis, we will adopt the center-of-mass of the system (which is the system center) as our unique perspective and will translate all measurements in other reference frames into this one. Thus, when an observer on our ring measures an angle to our "test" mass, we will realize that the "test" mass was at the relevant position at an earlier "time" as measured by this observer and at a still earlier time when information concerning this observer's measurement reaches our preferred frame -- the signal from the "test" mass took time to travel to the observer and the observer's communication of this information could travel no faster than a light signal from that observer to us. Nevertheless, the angle and therefore point of maximum elevation should be ascertainable and, because we are assuming that we have launched the "test" object at less than escape velocity, we should be able (assuming that our life span is essentially unlimited or that the experiment is conducted by generations of investigators) to make multiple measurements as our "test" objects repeatedly move out, slow, collapse in, collide elastically and again move back out.

As we begin our effort to measure distances along the central axis, we immediately encounter the first in a series of problems. Our "test" masses require kinetic energy to reach each of the distances we want to measure, and, to get this kinetic energy, we will need a finite amount of potential energy -- some finite quantity of fuel -- to make this happen. Because this potential energy has a mass equivalence, and because we have denied ourselves the power to increase the mass of our system without considering the impact of this mass increase on the overall problem, we will need a third mass at the system center to do our work separating the test masses. The amount of fuel required will depend on the maximum distance that we will want to mark along the central axis (assuming, of course, that we will not use enough fuel to exceed the escape velocity of the system), and, *significantly, on the magnitude of our test masses.* Thus, to be able to mark the distance that is a consequence of our "test" masses leaving the central point with a velocity of .95 percent of the speed of light, we will need fuel that, when added to the mass of the test objects is twice 3.20256 times each test body's rest mass (using the formula from the second of the concepts above)[95]. Note that this assumes that this fuel mass-equivalence is sufficiently low that its gravitational influence can be ignored. Since we will pick test masses of the lowest possible magnitude and we have picked a speed that results in a modest multiplier to this mass, we will initially assume that this is possible. To be able to mark the distance that is a consequence of our "test" masses leaving the central point with .995 percent of the speed of light, we will need fuel that, when added to the mass of the test objects is 10.01252 times each test body's rest mass. We must again assume that the mass of the fuel is insignificant in relation to the mass of the system. Of course, a second problem is

---

95   The formula by which this multiplier is calculated is the one from 5. b., above. What is contemplated by our example is that, after the start of the "experiment," there will be a sequence of elastic collisions between the two test particles, each returning to the origin at .95 percent of the speed of light under the influence of the gravitational pull of the system and each rebounding from the origin at that speed. The mechanics of such collisions are discussed in detail in W.G.V. Rosser, An Introduction to the Theory of Relativity, Butterworth & Co. (Publishers) LTD. (1964) at section 5.6 beginning on page 207.

developing. To each distance there corresponds a unique quantity of energy or fuel so that we must begin the exercise will a tank containing fuel with an adequate energy content to reach the farthest distances we intend to measure. Here again, at lease initially, we will assume that the test masses are sufficiently modest and the speeds reached sufficiently low that the required fuel is not too great a perturbing influence.

After we have calibrated distances along this central axis, we assume that objects that are twice the rest mass of our test mass are placed at the maximum height above and below the plane of our system reached by the test masses (i.e., at distance b and at distance c). As these objects accelerate toward the central point, existing thought -- based on the principal of equivalence -- suggests that each receives the same accelerations as our test objects received. Thus, each should arrive at the center point with the same velocity that our test objects left with. Of course, when these objects meet, the resulting system is more massive than our original system because the mass equivalence of the kinetic energy that these double-sized objects have accumulated is twice the mass-equivalence of the original fuel we used[96]. Accordingly, it is not possible to rigorously apply the principal of equivalence in this system. If the two doubled test-masses had been placed at the system center with the original amount of fuel used, they would not have reached the calibrated distances and if we put in the proper amount of fuel, we have violated the requirement above that "our model be a complete system and that no mass or energy pass its boundaries throughout each of our hypothetical experiments".

A critic might note, of course, that it was not possible to suddenly double the masses located along the central axis as we did in the example above and remain true to our requirement that we add nothing to our

---

96 See again, French, A.P., Special Relativity, W.W. Norton & Company, Inc. (1968), pages 172 to 175; and W.G.V. Rosser, An Introduction to the Theory of Relativity, Butterworth & Co. (Publishers) LTD. (1964) at section 5.6 beginning on page 207 and 5.7 beginning on page 217.

system. We cannot, after all, conjure these doubled objects out of nothing -- they must be moved into position from somewhere. Moreover, to get from where they were to the point at which we would start to observe them in earnest, we would need a complex series of steps and, significantly, each of these steps would require that still other objects be set in motion (to conserve momentum as well as energy). Further, the doubled objects and all of the other objects that would be disturbed in our effort to get them into position would have gravitational influences and the disturbances caused by our rearrangement would leave gravitational traces. In the end, then, our effort to get these doubled masses into position would leave us with a system to analyze that is different (although perhaps only slightly) than the one we started with.

Notwithstanding the validity of this criticism, the central point that we set out to demonstrate has indeed been established. After all, it is not that the example we have used is "invalid" while those before us have used "valid" examples, but, instead, that no one to date has made a valid, rigorous analysis of any gravitating system. Physicists have been taking liberties with the universe since almost the beginning of the organized study of physical systems. This is not, to be sure, a criticism of those that have taken such liberties. The organized study of physical systems would not have progressed to the point that it has reached if approximations had not been made and systems simplified. The complexity of the real universe is crippling. Nevertheless, there comes a time when systems that have been simplified and taken out of context must be reintegrated into a coherent whole and when approximations previously made must be re-examined to see if the effects ignored truly remain insignificant. The author believes that the simplifications that historically have been made in order to study simple mechanical and electro-magnetic phenomena fail us when analyzing large gravitating systems. In the future, physicists examining large gravitating systems will not be above to assume systems into existence or to assume that they can do experiments far from other gravitating bodies[97]. One must work with things that already exist and one must pay attention to conservation of momentum, mass/energy, angular momentum, etc., as one rearranges the components of the universe to illustrate any physical concept. The ultimate guiding principle is that there is not now nor can there ever be a place in the universe where the cumulative effect of all of the mass in the universe can be ignored[98].

---

97 The following quotation from Valens, E. G., The Attractive Universe: Gravity and the Shape of Space, The World Publishing Company (1969) starting at page 94 succinctly describes the problem:

> Our imagined experience on and around Eros should have made clear one fundamental and perhaps surprising fact about space: due to the nature of gravity, nothing can ever be at rest in space.
>
> This conclusion does not depend upon the fact that there is no such thing as "absolute" rest. There is no such thing as "relative" rest either. It is forever impossible to remain at rest in relation to any object or point in space unless you are physically in contact with such an object and are thus a part of it. Every thing is always moving in relation to any and every other thing.
>
> For example, you cannot "place" yourself at any point in space without giving yourself a definite velocity, and this velocity automatically binds you to one definite, unique orbit. Your velocity (speed or direction or both) changes at once and continues to change forever.

98 This, of course, is consistent with the spirit although not the actual formulation of what has been termed "Mach's principle," concerning the physical origin of inertia. According to Mach, "when the subway jerks, it's the fixed stars that throw you down." See Wikipedia, the free encyclopedia, the article entitled "Ernst Mach" and also the article entitled "Mach's Principle". In the author's view, when the subway jerks, the energy and force constraints imposed by the aggregate electric and magnetic fields from all of the charged particles in the universe (as translated into the frame of reference of the subway and then "you") throw "you" down. These fields are infinite in extent so that every charged particle in the universe (in its current position and with its instantaneously current motion as measured in the rest frame of first the subway and then "you,"

Another critic might assert that the increase in system potential energy implicit in the simple doubling of our test mass is so small and the rearrangement of the system necessary to bring these doubled masses into position is so inconsequential that we will not see a material impact

---

the falling rider) contributes some amount to the electric and magnetic field components and their energy at the point of observation in the rest frame of each. It might be protested, however, that all of the actors in the universe/subway system are "neutral" so that electrical forces can be discounted. But how can it be possible to know what is really neutral and what is not in a universe in which our test apparatus is inevitably compromised because affected by the very conditions we are trying to test? Consider, for example that, in common understanding, the Earth is an electrically neutral system but has a permanent magnetic field. If the Earth has a permanent magnetic field, however, it must be possible to find an almost infinite collection of frames of reference -- in fact, all frames but one -- in which the Earth has electric field components of some description. The electric fields in these alternative frames would induce polarizations in the materials of which the Earth is made in these alternative reference frames. Thus, it may simply be that the sudden disparity of the Earth's field measurements and the resulting induction effects between the frame of the accelerating subway car and the frame of the stationary "you" -- relying on the Lorentz Transformation equations appropriate to each frame -- accounts for the fall. Further in this regard, consider that, although every physicist knows that no electric field may exist inside a hollow conductor or Faraday cage, the author believes a magnetic field penetrates that hollow conductor and, therefore, that a compass would work inside such a cage. Since the magnetic field that influences the compass has electric field components in all reference frames save the one in which the compass is stationary, there are some frames of reference in which an "electric" field would be found inside such a hollow conductor. Moreover, the "particles" of electromagnetic radiation -- photons -- and the radiation itself are generally considered to be "neutral." This notwithstanding, when these "neutral" moieties are incident on typical matter, they separate electrons from the much more massive protons in that traditional matter and thereby induce local electric fields. It is the author's view that the focus on separate electric and magnetic fields that is a relic of physicists' discovery of the phenomena have left all of us without a clear conception of just how related these phenomena are.

on the behavior of the system. In many cases, we likely would not be able to measure it -- twice an inconsequential amount of fuel added to a system with an escape velocity of .95 percent of the speed of light would not allow the demonstration of a significant deviation from the equivalence principal. If we double and redouble and redouble again and again the test mass, however, sooner or later the mass equivalence of the fuel required to send this object mass on its journey will become a material fraction of the overall system mass. At this point, it is no longer appropriate to ignore the influence of the fuel on the behavior of the system and, of course, the system is no longer the planar ring that we started with. Further, no matter how inconsequential the test mass we start with, as the system escape velocity tends toward the speed of light, the potential energy required to send the test object to the point just short of "escape" tends to infinity and, again, this potential energy will dictate system behavior. Our example does, accordingly, establish the failure of the equivalence principle.

Ultimately, then, just as classical mechanics fails when dealing with objects that have velocities approaching the speed of light, the principal of equivalence fails when dealing with many aspects of systems that have escape velocities approaching the speed of light.

A determined critic, to be sure, would nevertheless insist that the discussion above absolutely must be wrong because there have been incredibly precise time-of-fall experiments done on the earth that support the principle of equivalence as it is presently understood[99]. The

99 It has been suggested that the principle of equivalence addresses two discrete questions. First, do objects of different mass fall at the same rate and, second, do objects of different composition (but not necessarily different mass) fall at the same rate? See, G. Tauber, Albert Einstein's Theory of General Relativity, Crown Publishers, Inc. (New York, 1979), Part III which begins on page 114. An appropriate test of the second principle is to drop objects of differing compositions from the same point and measure the time-of-fall and there is some suggestion that such tests have been performed. The truth, however, is that they have not.

The experiments that have been done carry hidden (and erroneous) assumptions that are ancient in origin -- none of us, the author certainly included, has yet fully come to grips with the picture of the universe that 21st century physics forces on us. In particular, although we routinely assume otherwise, it is now clear that any gravitational experiment that we may conduct on the Earth cannot really be "repeated" because the location of the Earth in the universal system is constantly changing due to the Earth's rotation, its revolution around the sun, the sun's revolution around the galactic center and the galaxy's motion with respect to the rest of the matter and energy in the universe. Thus, while ancient lore suggests that Galileo or his contemporaries took masses to the top of a tall tower on the Earth and measured the time it took for each to fall to the ground and did so repeatedly and we might perform these experiments again today with sophisticated timing devices and other sophisticated technology to "refine" these measurements, we have really not accomplished what we set out to do. It is now clear that we have only appeared to measured the time it takes objects to fall between two "fixed" points. If the earth is dethroned as the center of the universe, there is no escaping the fact that the earth must move through the universe, and, as the earth travels through the universe, the apparently fixed point on the tower and the fixed point on the ground are, in reality, in rapid motion and two objects dropped from the top of the tower even seconds after each other really fall from and to points that are unique and different and perhaps thousands of miles away from each other. On the other hand, after the earth loses its place at the center of the universe, it becomes all the more remarkable that we can repeatedly take objects on apparent cycles through gravitational "fields" and expect gravitational potential energy to be conserved -- given what we now know, the fact that gravity may be treated as a central force with changes in gravitational potential energy determined to be path independent is absolutely staggering for no "cycle" that has been analyzed has every really been a "cycle." Instead, each object taken through an apparent cycle has simply moved along a complex path and that path only appears to have been a "cycle" when measurements are made with respect to systems in the vicinity of the object and on similar paths. The author will provide his detailed explanation how gravitation can cause central force like behavior without being a force at all in a later section of this work. At this point, the author would simply note that all systems that are currently known to exist are electrical in origin so that the fact that energy conservation constrains changes in electrical and

discussion above, however, indicates that there is no time-of-fall experiment that could be done on the earth that would reveal the failure of the equivalence principle because the fields available and distances involved are not adequate for gravitational potential energy to have a measurable impact[100]. Would these critics reject the now well-established fact that rapidly moving objects have greater inertial mass than "identical" objects at rest simply because the mechanics of Special Relativity cannot be proven by experiments that involve billiard balls and the speeds encountered in a typical game of pool?[101]

---

magnetic fields would be expected to produce affects on all systems that mimic what has previously been described as gravitational behavior.

100 The experiments of Eotvos and Dicke, which are different in kind from the ancient and modern "time of fall" experiments, admittedly require a different analysis to determine whether the author's ideas can be made to fit within the constraints they impose. The author believes that his central concept -- that the energy content of an object is tied to the electro-magnetic signatures of its constituents for all purposes -- would be consistent with the requirements of these experiments. Gravitational mass and inertial mass of all objects are the same because all "mass" is the result of the energy content of the fields that define the objects. Further, two objects behave in a similar manner when one is substituted for the other because the behavior of the electrodynamic components of both are dictated by the positions of all of the other electrodynamic components in the universe. So long as the positions of both with respect to the other components is similar and so long as neither object represents a material percentage of the substance of the universe, their behaviors will be similar and they will fall at the same rate and respond to forces with the same resistence.

101 Relativistic pool is a radically different game than that really played. Compare, Rindler, W. Essential Relativity, at Section 2.3 on pages 26 to 27 with Section 5.8 on pages 86 to 87. (Springer-Verlag 1977).

9. Further Analysis of the Foundations of Einstein's General Relativity Theory

a. Problems with Foundation Experiments

The discussion above purports to establish that the "Principal of Equivalence" that is the central tenet of Einstein's theory of General Relativity is mistaken. Earlier sections have also challenged other assumptions that are currently part of the lore of General Relativity. Specifically, earlier sections have called into question the validity of the concept of a gravitational "field", the concept of a "test" mass that may be moved about within such a "field" and the assumption that the Newtonian model for gravitation can be used so long as one is far from objects with relativistic escape velocities. Ultimately, then, there must be problems with all of the foundation experiments on which General Relativity is based. In this section, we will explore these experiments and endeavor to find these flaws. The discussion above suggests that the re-examination should look for aspects of the foundation "experiments" -- which, after all, are almost entirely "thought" experiments -- which ignore or mis-measure the gravitational impact of some form of potential energy operating during each "experiment" and this will be our underlying theme.

Let us begin this further discussion by introducing and challenging the premise of perhaps the most basic of the foundation experiments of General Relativity -- a "thought experiment" involving an enclosed chest or enclosed elevator car or rocket ship without windows. The following description, prepared by Albert Einstein himself[102], describes the situation:

> We imagine a large portion of empty space, so far removed from stars and other appreciable masses, that we have before us approximately the conditions required by the fundamental law of Galilei. It is then possible to choose a Galilean reference-body for this part of space (world), relative to which points at rest remain at rest and points in motion continue permanently in uniform rectilinear motion. As reference-body let us imagine a spacious chest resembling a room with an observer inside who is equipped with apparatus. Gravitation naturally does not exist for this observer. He must fasten himself with strings to the floor, otherwise the slightest impact against the floor will cause him to rise slowly towards the 'ceiling' of the room.

> To the middle of the lid of the chest is fixed externally a hook with rope attached, and now a "being" (what kind of being is immaterial to us) begins pulling at this with a constant force. The chest together with the observer then begin to move 'upwards' with a uniformly accelerated motion. In course of time their velocity will reach unheard-of values -- provided that we are viewing all this from another reference-body which is not being pulled with a rope.

> But how does the man in the chest regard the process? The acceleration of the chest will be transmitted to him by the reaction of the floor of the chest. He must therefore take up this pressure by means of his legs if he does not wish to be laid

---

102 See G. Tauber, Albert Einstein's Theory of General Relativity, Crown Publishers, Inc. (New York, 1979), Section C., at pages 75 to 76 with the italics in the original.

out full length on the floor. He is then standing in the chest in exactly the same way as anyone stands in a room of a house on our earth. If he releases a body which he previously had in his hand, the acceleration of the chest will no longer be transmitted to this body, and for this reason the body will approach the floor of the chest with an accelerated relative motion. The observer will further convince himself *that the acceleration of the body towards the floor of the chest is always of the same magnitude, whatever kind of body he may happen to use for the experiment.*

Relying on his knowledge of the gravitational field (as it was discussed in the preceding section[103]), the man in the chest will thus come to the conclusion that he and the chest are in a gravitational field which is constant with regard to time. Of course, he will be puzzled for a moment as to why the chest does not fall in this gravitational field. Just then, however, he discovers the hook in the middle of the lid of the chest and the rope which is attached to it, and he consequently comes to the conclusion that the chest is suspended at rest in the gravitational field.

Ought we to smile at the man and say that he errs in his conclusion? I do not believe we ought to if we wish to remain consistent; we must rather admit that his mode of grasping the situation violates neither reason nor known mechanical laws. Even though it is being accelerated with respect to the "Galilean space" first considered, we can nevertheless regard the chest as being at rest. We have thus good grounds for extending the principle of relativity to include bodies of reference which are accelerated with respect to each other, and as a result we have gained a powerful argument for a generalization of the postulate of relativity.

---

103 For those interested in reviewing this discussion, the reference is G. Tauber, *Albert Einstein's Theory of General Relativity*, Crown Publishers, Inc. (New York, 1979), Section B., at pages 74 to 75.

We must note carefully that the possibility of this mode of interpretation rests on the fundamental property of the gravitational field of giving all bodies the same acceleration, or, what comes to the same thing, on the law of the equality of inertial and gravitational mass. If this natural law did not exist, the man in the accelerated chest would not be able to interpret the behavior of the bodies around him on the supposition of a gravitational field, and he would not be justified on the grounds of experience in supposing his reference body to be "at rest."

The first thing to note regarding the discussion above is that it relies on the intervention of a "being" that Einstein has excused from complying with the basics of mechanics either as Newton understood them or as adjusted to conform to the requirements of Special Relativity. Specifically, this "being" is able to produce an acceleration of the "chest" without apparently being obliged to expend energy and transfer momentum -- in this example, the "being" is able to generate velocities for the chest and its contents of "unheard-of values" without the use of a source of potential energy and without accelerating some corresponding objects to "unheard-of values" in the opposition direction. In this discussion, then, Einstein has relied on the miraculous.

Similar discussions, of course, can be found in many texts by other authors and most avoid the need to rely on such obvious sleight of hand. Thus, in most texts, the chest becomes an "elevator" or a "rocket ship." While either vehicle has a greater veneer of plausibility, careful analysis leads to the conclusion that this veneer only serves to conceal fatal flaws in the underlying logic.

Take, for example, the use of an elevator. A typical elevator consists of a motor, a distant pulley, a chest or car in which to ride and an extended cable strung from the motor, over the pulley and attached to the chest or car. Those that design elevator systems would surely balk at any requirement that a single motor of modest power be used to give the same acceleration to objects of widely different masses, would insist that the cable used be strong enough to transfer the accelerations to be transferred and would consider the mass of the cable (which depends on its strength and also its length) in determining the load the motor would be required to handle. Further, an elevator system of trivial mass could not be used to accelerate objects of great mass. If such were attempted, it would be the elevator system and not the elevator car that was accelerated. Thus, the motor and the pulley would need to be moored to systems that are always substantially greater in mass than the object in the car. Therefore, if the object in the car had considerable mass, then the motor with moorings, the motor's energy source, the pulley with its moorings and the cable would all be gravitating bodies of even more considerable mass and these massive objects would inevitably provide a framework of gravitating things with respect to which those in the chest or car could fix their location. To be sure, one could attempt to move these bodies so far away from the chest or car in an effort to eliminate their gravitational influence, yet, in doing so, one would need a very long and very strong (and very massive) cable. If one imagined putting a hypermassive object in the car and designing a cable that has adequate tensile strength to "pull" on such an object and the full length of the cable, a motor with moorings adequate to give such an object a constant acceleration (rather than simply reeling the motor and moorings in) and a hypermassive pulley with moorings over which to run this cable, the veneer of plausibility evaporates.[104] Such a system simply could not exist without providing a universal framework against which motion could be measured.

Of course, there remains the frequently used device of generating the required acceleration with a rocket ship. Even this device has a problem -- a problem that ultimately can be traced back to the misunderstanding of the gravitational influence of potential energy that has

---

104 How, in fact, could one put a "black" hole -- which apparently is assumed to be a singularity of no real spacial expanse -- into such an elevator? What would happen to the cable as one tried to attach it to such an object?

been discussed previously. Whenever one reads a book discussing Relativity and whenever one sees an analysis of Relativity principles, one will always note that none of the "rockets[105]" of Relativity -- Special

---

105 Although the foundation materials regarding Relativity frequently discuss the behavior of "rockets," the concept of a "rocket" used is, at best, an anachronism. As we in the 21st century can testify, a rocket is a system, not an object. It is a device for the conversion of potential energy into kinetic energy which typically transfers a portion of this energy (and provides momentum in a desired direction) to a subpart of the overall system -- the payload -- so that the payload can be moved toward a desired location. Inevitably, however, to provide this energy and momentum to the payload we are required to impart kinetic energy and momentum to less relevant portions of the same system and so we release plumes of exhaust in the equal and opposite direction from which the payload is to be moved. Significantly, absent the operation of an external force, the center-of-mass of a real "rocket" system cannot be moved. See, e.g., D. Tilley, University Physics for Science and Engineering, Cummings Publishing Company, Inc. (Menlo Park, California 1976), Section 6.2 at pages 84 and 85. The author would further note as an aside his suspicion that, if one interviewed the rocket designers of the world, past and present, they would add an additional requirement to the concept of a rocket that strikes at the very heart of this discussion of the principle of relativity. A real rocket has a *guidance* system so that we can send our payload to the unique places in the universe that we want to explore. The very concept of a guidance system, however, is contrary to the central tenet of relativity -- that all frames of reference are equivalent. Quite the contrary, the concept of a "guidance" system implies (and it is an unquestioned but nevertheless unconscious assumption in astronomy) that all frames of reference in the universe are unique. It is for this reason that the teams of scientists that have created probes to search beyond our solar system have never attempted to build a rocket with the power required to accomplish their missions along all trajectories possible. Instead, they have built much much smaller rocket systems but have used complex guidance systems to allow their probes to follow precisely chosen paths so that they can steal momentum from the larger objects in the Solar System. The actions taken by the legions of sophisticated experts handling the real rockets and payloads that have been launched within and beyond our Solar System belie the conceptions of rockets

or General -- ever contain any "fuel" and, thus, the gravitational "fields" associated with the fuel that must be in these rockets if the rockets are to cause the accelerations that are required -- even if it is an ideal fuel that can be entirely converted into kinetic energy -- are systematically ignored. If one puts real fuel -- a fuel that exerts a gravitational pull consonant with work it will ultimately be called upon to do -- in the rockets of Relativity, the outcomes of the various thought experiments change and, at high "field" strengths pulling on very massive objects, change dramatically. A critical reason for the change is that the strength of the gravitational "field" associated with this potential energy is inextricably tied *both* to the amount of acceleration to be imparted *and* to the mass of the object or objects to be accelerated.[106]

Thus, if one starts with a given payload mass, given gravitational "field" strength to replicate and given final velocity, one can calculate the energy content of the fuel required to generate the target velocity (and thus target kinetic energy) and presumably one can expend this fuel rapidly enough to produce a constant acceleration equal to the gravitational acceleration one is trying to mimic. So far in this process, however, there has been no consideration of the gravitational interaction between the payload and the energy to be imparted -- the potential energy in the fuel one must put in the rocket. Thus, if one assumes that one has imparted an acceleration of precisely g_{Model} to a moderate mass

and observers that are central to Relativity. Indeed, the utility of the gyroscopes that are the core of every "inertial" guidance system only serves to prove the original wisdom of Isaac Newton and his intuition that rotation proves the existence of an absolute *mechanical* frame of reference. The ability to navigate using such a guidance system, of course, is substantial evidence that there is an absolute frame of reference as Newton advocated. The author will attempt to rehabilitate Newton's belief in the concept of an absolute time further below.

106 This, of course, is diametrically opposed to the belief that all objects fall in a given gravitational "field" at the same rate. The discussion earlier, however, established that this is true only of weak filed over modest distances and affirmatively is not true of strong ones.

of say $M_{Payload}$, by expending a fuel with the potential energy content of the final kinetic energy amount to be reached, the actual acceleration perceived by $M_{Payload}$ would be g_{Model} *less* the acceleration caused by the force on the payload due to the gravitational interaction with the fuel that will launch it. Initially this is $G(M_{Payload})(M_{Fuel\ Launch\ Object\ 1})/d^2$ divided by $M_{Payload}$, with d being the distance between the payload's center-of-mass and the fuel's center-of-mass and the opening "G" in the formula being the gravitational constant. A fundamental constraint on this system is provided by the famous equation $E = mc^2$. Thus, the final kinetic energy of the launched object divided by c^2 can be no more than the mass of the launch fuel[107] so that, for a launch to a non-relativistic speed:

$$(M_{Fuel\ Launch\ Object1}) > \tfrac{1}{2} M_{Payload}(V_{Final\ for\ Payload})^2 /c^2$$

The actual initial acceleration of our payload, then, would be no more than:

$$g_{Model} - GM_{Fuel\ Launch\ Object\ 1}/d^2$$

or

$$g_{Model} - G\tfrac{1}{2}M_{Payload}(V_{Final\ for\ Payload})^2 /c^2d^2$$

Significantly, then, our attempt to replicate a specific gravitational "field" strength has required the incorporation -- at least superficially[108] -- of a term that depends on the mass of the object to be accel-

---

107 The mass of the fuel would, in fact, need to be greater because our rocket's exhaust would need to carry momentum equal to the momentum of our payload and our fuel would need to do work on both our payload and its exhaust in order to allow this to occur.

108 The mass of the required fuel, of course, constrains the ultimate kinetic energy of the payload rather than the instantaneous acceleration. Accordingly, it would appear that one can replicate a particular

erated. Note as well that, if one dropped a second payload equal in mass to the original payload into the same system and also added the fuel required to accelerate this object to the same final non-relativistic velocity, the resulting acceleration would be still less than the target figure. The actual acceleration of an object in this system would be:

$$g_{Model} - G[(M_{Fuel\ Launch\ Object\ 1}) + (M_{Fuel\ Launch\ Object\ 2})]/d^2$$

As before, the requirement that the final kinetic energy have a mass-equivalence no less than the mass of the fuel expended means that the mass of the fuel for this experiment would need to meet the following condition:

$$(M_{Fuel\ Launch\ 2}) > (M_{Fuel\ Launch\ 1}) + \tfrac{1}{2}\ (M_{Payload})\ (V_{Final\ for\ Payload})^2/c^2$$

or

$$(M_{Fuel\ Launch\ 2}) > \tfrac{1}{2}(M_{Payload})\ (V_{Final\ for\ Payload})^2/c^2 + \tfrac{1}{2}\ (M_{Payload})\ (V_{Final\ for\ Payload})^2/c^2$$

or

$$(M_{Fuel\ Launch\ 2}) > (M_{Payload})\ (V_{Final\ for\ Payload})^2/c^2$$

The author suspects, of course, that, in many instances, just enough fuel could be added (and its use carefully metered) to increase the acceleration of all of the objects in our rocket to the point that we could actually reach exactly g_{Model} and achieve our final velocity so long as g_{Model} and our payload are not too large. Our model, then, would

acceleration for objects of a variety of masses with a fixed quantity of fuel simply by assuming that the acceleration occurred for an ever diminishing period of time. As the mass to be accelerated tends to infinity, the period of acceleration would tend to 0. There are alleged proofs of the validity of General Relativity that appear to rely on this feature of the mathematics to establish that the formulas of Special Relativity can always be used in some vicinity of points in space-time.

replicate, but only for a limited number of objects, with a lower limit on the ultimate final velocity, and likely with a unique metering signature of gravitational subtleties, the gravitational "field" of M_{Model}.

In other instances, however, it seems unlikely that we could replicate the gravitational "field" of our proposed model if we are dealing with the acceleration due to an extremely massive object and if we are attempting to accelerate another extremely massive object at the target rate of acceleration. Indeed, if we are given the gravitational "field" at the surface of an extremely massive object to model (say an object with a surface pull of ½ of that at the Schwartzchild radius of a "black" "hole") and if we assume that we want to accelerate an object that also has a surface pull of 1/2 that at the Schwartzchild radius of a "black" "hole" (or 3/4 or 7/8 or 15/16, etc.) away from this point, can any quantity of fuel be inserted to produce the target acceleration? Could we produce any acceleration at all?

Although the author would acknowledge a lack of rigor in the discussion above -- the author has only begun to outline a new and complex formula to measure the force between two objects that are being separated from each other by the use of a perfect fuel -- nevertheless, intuitively, the conclusion seems inevitable that, given any object with a finite rest mass, there will be a maximum kinetic energy and thus maximum distance allowable[109]. Beyond this energy and distance, the

---

109 To add another thread to the argument above, the author would note that one familiar consequence of Special Relativity is that no object with a rest mass may be accelerated to the "speed of light" and that the maximum velocity for the transmission of energy or momentum is again "light speed". Therefore, the optimum rocket fuel is one that can be converted entirely to electromagnetic radiation and the maximum exhaust velocity of any fuel is the speed of such radiation and thus the "speed of light." Because there is a cap to any rocket's exhaust velocity, one can increase the momentum to be imparted to a payload only by increasing the quantity and therefore the mass of fuel to be used in launching that payload. For a more rigorous analysis of the relationship between the mass of a rocket's fuel and its payload the reader is referred to the

gravitational pull of the additional fuel that is hoped to cause still greater acceleration, speed, and distance actually overcomes the effect of the expenditure of this fuel and the real distance is less. Beyond the maximum distance, additional fuel inevitably becomes less and less potent until the point at which this extra fuel is, in fact, impotent. The object cannot accelerate away from the fuel/object assembly and the whole system collapses.[110]

b.   Problems with Missing Pulses

In addition to concerns with the so far ignored gravitational interaction between rockets and payloads in Relativity's foundation experiments,

---

discussion of the "Photon Rocket" at French, A.P., Special Relativity, W.W. Norton & Company, Inc. (1968), pages 183-184. As noted in that discussion the fraction "f" that a payload may be of the total mass of the best conceivable rocket -- a rocket with the optimum exhaust velocity of electromagnetic radiation -- is a function of the maximum speed of the rocket's payload. Therefore, an increase in payload mass necessitates an increased fuel mass. Since each unit of payload mass will interact with all of the units of fuel mass, there is an ever increasing gravitational interaction between payload and fuel -- the increase in momentum grows arithmetically with increasing fuel but the gravitational interaction grows exponentially. Thus, for example, if one doubles the payload and then doubles the fuel in the launch vehicle to attempt to achieve the same final momentum per unit mass, the gravitational interaction between fuel and payload will be 4 times as great. Further, if one doubles the payload and fuel again, the gravitational interaction will be 16 times as great and so forth. In addition, if one takes a nominal mass and doubles and redoubles and redoubles it without limit and if one takes what physics currently calls a "black hole" and halves it and rehalves it again and again without limit, sooner or later the result of the first exercise will have a mass that exceeds the result of the second exercise.

110 This, of course, is contrary to the now popular conception that, given sufficient fuel, one can accelerate all objects to speeds that approach ever closer to (but can never exceed) the speed of light.

the author also has concerns with the rigorous application of one of the most basic of the assumptions made in the course of the foundation thought experiments of General Relativity -- that the cycles of electro-magnetic radiation are an important " invariant" that can be relied on for the measurement of the passage of time in a reference frame. In the most familiar and most basic formulation of one of the core concepts of General Relativity[111], there is a rocket ship with two "clocks", one in the back of the ship and one in the front. Just as the rocket begins to accel-erate with a constant acceleration of "a", a series of pulses is sent from the clock in the rear of the rocket toward the clock in the front. Because the clock in the front is accelerating away from the clock in the rear, the clock in the front receives "pulses" less frequently and, thus, the determination is made that accelerating clocks tick more slowly than stationary ones.

This analysis, coupled with the principle of equivalence, led Albert Einstein to believe that clocks at a higher gravitational potential had to tick more quickly than clocks in a lower gravitational potential and thus that the passage of time is affected by the local gravitational po-tential. He reasoned that the number of "pulses" was a fixed number and that a gravitational "field" has the same impact as the acceleration in the example above because constant acceleration and the impact of a gravitational "field" were, in his view, indistinguishable. The author believes that there is an imprecision in this analysis that was barely noticeable to physicists at the end of the 19 century and the beginning of the 20th but that is glaring today. Specifically, which is the invariant quantity associated with a "pulse" of "light"? Is it the detectable cycles of the interlinked electric and magnetic fields associated with a signal or is it the number of photons[112] that are included in the signal? The

---

111 See W.G.V. Rosser, An Introduction to the Theory of Relativity,
    Butterworth & Co. (Publishers) LTD. (1964) at Section 12.3 beginning
    on page 441.

112 An additional concern of the author is whether it is possible to convert a radio
    wave to a gamma ray and vice versa by changes of coordinate systems.

former are electromagnetic phenomenon and thus it should not be possible to identify a preferred frame of reference by analyzing them. On the other hand, the latter are mechanical phenomena and, if the arguments above and that follow are valid, it should be possible to locate a preferred frame with the assistance of clocks that measure the arrival of photons rather than with the counting of cycles of electromagnetic radiation[113]. If we adopt a standard atomic clock (as the author believes has been done as the current standard clock) and circulate the standard, can we not use the number of photons received, regardless of wavelength, as a measure of the passage of time? Further, if there is a standard, cannot we measure the particular "red" or "blue" shift of the radiation from a particular source and then compute a maximum relative gravitational potential and a maximum potential relative velocity for the source of the radiation.[114] Couldn't we also develop an integrated view of the universe that would allow us to separate the shift attributable to both of these sources? We will discuss this further below.

c.   Problems Locating an "Inertial Frame" Anywhere

In addition to concerns associated with ignoring the mass of the fuel required to accelerate objects in the foundation experiments of General Relativity and the proper identification of invariants connected with electromagnetic signals, the author also has problems with the idea that it is possible to apply the formulas of Special Relativity rigorously at

---

113 Although the author is not certain, he thinks that the argument he is making is similar to the argument of Brillouin in Chapter 3 of his work, Relativity Reexamined. See L Brillouin, Relativity Reexamined, Academic Press (1970), especially the discussion in Section 4 beginning at page 33.

114 Hasn't this, in fact, been done as evidenced by experiments which rely on the "Mossbauer Effect." See W.G.V. Rosser, An Introduction to the Theory of Relativity, Butterworth & Co. (Publishers) LTD. (1964) in Appendix 6 beginning at page 493.

any single point (much less over any finite region of space) anywhere. In particular, the author disagrees with the established lore of Relativity that it is always possible to replace a frame of reference in a gravitational "field" (of whatever intensity) with an equivalent uniformly accelerating frame and that, so long as the area of space involved is sufficiently small, an observer in the enclosed space cannot distinguish between the accelerating frame and a frame at rest in a gravitational "field"[115].

To illustrate the concern, imagine a frame of reference in which there is but a single electron and that this electron is far from any gravitating mass. Most of the texts the author has read suggest that Special Relativity could appropriately be used in the "vicinity" of this single electron because the gravitational "field" of this electron is "inherently" weak. This notwithstanding, there is a frame of reference permissible in Special Relativity in which an observer could be adjacent to this electron and assign this electron the mass of a house, another such frame in which the electron would be assigned the mass of the earth, another in which the electron would be assigned the mass of the sun, yet another in which the electron would be assigned the mass of a neutron star and still others in which the assigned mass of the electron is greater and more fantastic still. Objects in the vicinity of this electron cannot all behave in a manner consistent with all of these alterative masses and this electron therefore cannot curve space-time in order to make its gravitational presence known.

In the same vein, if one examines the collision of two billiard balls on a pool table from the point of view of an electron moving toward the collision point with a straight line trajectory perpendicular to the table and that includes the collision point, one can make the masses of the billiard balls as great as one pleases simply by envisioning the electron moving at greater and greater speeds. Indeed, there must be some speeds of the electron in which the billiard balls would be so massive

---

115 See, M. Born, Einstein's Theory of Relativity, Dover Publications, Inc. (1965) at 357.

that they could not rebound after the collision, yet an overall system that consists of two billiard balls does not behave as a collapse into a "black hole" in at least some frames of reference. There are some frames of reference in which a game of pool is only a game of pool. Accordingly, those reference frames that see the collision of these two pool balls as the collapse of two objects into a black hole cannot be real and the laws of physics are inherently different in a real spectator's frame as compared to these frames.

The conclusion is therefore inescapable that some of the frames of reference that are permissible in Special Relativity are not permissible in the real world[116]. Indeed, if one installed a series of clocks and observers all with nominal mass in one reference frame, these clocks and observers would, nevertheless, be so massive when viewed from other purportedly equally valid reference frames that the interaction between the two would prevent either from being considered an inertial frame. Obviously, once one installs any object with a rest mass in one frame of reference, some frames moving at some relative speeds could not

---

116 A frequent statement in many texts on Special Relativity is that it is always possible to adopt a frame of reference in which a "proper" time measurement can be made and, conversely, it is always possible to adopt a frame of reference in which a "proper" distance measurement can be made. See Tolman, Richard C., Relativity Thermodynamics and Cosmology, Oxford at the Claredon Press (1958), Section 18, pages 33-34 and W.G.V. Rosser, An Introduction to the Theory of Relativity, Butterworth & Co. (Publishers) LTD. (1964) at sections 3.10 beginning on page 125 with special emphasis on pages 127 to 128 and 6.4 beginning on page 266 with special emphasis on page 267. On the other hand, even if it were possible to convert the entire mass of the earth to kinetic energy in an instant and to send the smallest fundamental component of the earth out in any direction, there would be a limit to the proper distance and proper time measurements that the resulting observer could make. Limits on access to energy therefore prevent any one specific observer from having access to all frames of reference that are permissible in Special Relativity.

"co-exist" with the first frame and would collapse.[117] One frame or the other, then, must be unreal, and, if one finds oneself in the first frame, one would know that it is the second frame that is the unreal one.

The examples above inexorably lead to the conclusion that the kinematics of Special Relativity are at best only approximations everywhere.

d.   The Location of Absolute Positions in Space and Time

At this point, the author believes it is appropriate to address the relativity principal itself. The point of departure for this analysis is the now famous example of a space that purports to include a fluid spherical object and a fluid ellipsoid of revolution that are separated from each other at great distance and rotating with respect to one another.[118] According to the various books the author has read, each observer on each object surveys their respective home and one determines that his home is the ellipsoid of revolution and the other determines that his home is the sphere. The question is then asked: "What is the reason for this difference between the two bodies?"[119] Newton is alleged to have ascribed the difference to the rotation of the ellipsoid with respect to "absolute"

---

117 This is all the more certainly the case if one factors in the gravitational potential energy that is implicit in any extended system of clocks and observers as suggested in the opening sections of this work. Further, while it is possible to envision two coordinate systems moving at different rates with respect to each other at a particular point in space, it seems a stretch to envision both systems as co-extensive with a finite universe.

118 See A. Einstein, The Foundation of the General Theory of Relativity, §2, reprinted in The Principle of Relativity, Dover Publications, Inc.(1952), at 112 to 113. See also, M. Born, Einstein's Theory of Relativity, Dover Publications, Inc. (1965) at 309-312

119 See A. Einstein, The Foundation of the General Theory of Relativity, §2, reprinted in The Principle of Relativity, Dover Publications, Inc.(1952), at 112, bottom of the page.

space (and the author would agree with him if the fluid of which each objects is made is the same, the amount of fluid is the same and the objects are at the same temperature). Einstein, Mach and Born, of course, all find this unsatisfactory. The comments of Einstein, Mach and Born, however, are all, in their own way, unsatisfactory because each fails to consider whether it is possible to construct their hypothetical system in a real space and, in the construction process, also hide all traces of the construction work and thereby avoid the creation of the unique reference frame they chide Newton for using.

If one were assigned the task of assembling such a system in a space that has no preferred direction, in which mass/energy is conserved, in which mass/energy exerts a force active throughout all of space and in which momentum and angular momentum are conserved, the author believes that one would find the construction task difficult (in fact, as discussed further below, impossible if the requirement that the objects be identical is rigorously enforced) and, in any case, one would certainly find it impossible to conceal the work. Let us assume, however, that one would attempt the exercise.

Initially, one would need to take the two final objects and place them together at the center-of-mass of the final system. Given the assumption, however, that each droplet is made of the same substance and that the substance is a fluid, it seems reasonable that this double assembly would collapse into a single spherical droplet with its center located at the center-of-mass of the final system or at least that one would have to begin with such a droplet to keep true to our original assumptions. Otherwise one would have introduced a preferred direction in space -- that of the line between the centers of the two objects -- into the assembly.

If the universe contained only the spherical proto-dual object system, of course, it would forever remain the same and would forever be found at the center-of-mass of that universe. One therefore must introduce some form of potential energy -- some perfect fuel -- to use to perform work on this system in order to accomplish the task. This ideal fuel would be used to separate and reshape the fluid against the force of gravity into two "identical" spheres; to separate these spheres against the force of their mutual attraction (and the attraction of the fuel being used to separate them); to deform one of them; and to set one or the other of these objects in rotation. Given the discussion above which indicates that every quantity of work that is performed against the pull of gravity leaves a trace on the space where that work was done and that this trace exerts a gravitational influence throughout all of space, then, no matter how skillfully and efficiently the work is performed, it would not be possible to create such a system without leaving traces that are detectable by a measurement of the gravitational "fields" found throughout all of space. To know the total history of the construction enterprise, one need only examine the gravitational "fields" of the resulting system to locate traces of:

- the potential energy stored in the gravitational "fields" within this universe left in the process of reshaping the initial single sphere into two smaller spheres as well as the masses (the fuel's exhaust photons if one considers the potential energy in the form of a perfect fuel) that were required to be set into motion to allow conservation of momentum during the reshaping and separation process;

- the potential energy stored in gravitational "fields" of these two objects resulting from the expenditure of fuel that allowed these two objects to be separated from each other (and from the center-of-mass of the entire system);

- the potential energy stored in the gravitational "fields" of this system left by the apparatus that deformed one of the objects out of its spherical shape;

- the third object or the set of objects that at one time was adjacent to at least one of the ellipsoid/sphere set and that carries the remaining momentum and angular momentum to insure that the complete system has no net momentum and no net angular momentum; and

- the potential energy stored in the gravitational "fields" of the entire system that carried the third object or the set of objects away from whichever of the other two is "really" rotating with respect to the center-of-mass of the entire assemblage of objects and fuel.

Any attempt to extinguish these traces can never succeed. All that can be done, even in a larger system that contains other materials, is to dilute the traces of these indelible changes.

It might be protested, of course, that we have left open the prospect that the spherical object is the one that is "really" rotating. If this is what is discovered, however, it would be proper to conclude that one of the original assumptions -- that the two objects were identical -- is faulty.[120]

---

120 The author would add that he doesn't think that it is necessary to search all of space to prove conclusively that the objects cannot be identical as originally supposed. If the two objects are to be identical, then they should both exert the same gravitational pull on each other and through out all of space; they should contain the same number of discrete components and those components should be at the same temperature. If one of the objects is an ellipsoid of revolution, however, some work would have to have been done to deform this object from the normal spherical shape of a fluid that is not subject to any external forces. The potential energy that would be released if this object is allowed to relax to its low-energy spherical state inevitably has a mass equivalence, however, and, therefore, if both objects have the same gravitational mass, then the ellipsoid of revolution must have fewer components or may have the same components but at a lower temperature. The objects cannot be rigorously "identical".

e. The Unique Role of the Center of Universal Mass in Assigning Absolute Positions in Space and Time

Generalizing the discussion in the section above to a universal system, let the author suggest that it should always be possible to create a unique coordinate system in any universe that incorporates the concepts of conservation of mass/energy, conservation of momentum, conservation of angular momentum and a force that appears to act between mass/energy in proportion to the amount of work historically done in a particular volume of space. The unique origin of such a system would be its center-of-mass. Although this certainly runs counter to the existing understanding of Relativity, it is important to remember that, in this existing understanding, there are only three processes that can influence the running of "clocks" -- gravitational "fields", constant relative motion and accelerating relative motion. If one is to prevent repetitive physical processes coincident with the center-of-mass of a universal system from keeping perfect time,[121] then one would have to be able to create a gravitational "field" at this point or cause this center-of-mass to move. The very definition of the "center" of mass of a system that has no preferred original direction, however, dooms this enterprise.[122] Accordingly, while one certainly could force a particular physical apparatus that happened to be at the center-of-mass of a universal system into motion, the unique point of space that is

---

121 Because we have assumed that there is no preferred direction in our universe, the center-of-mass must be at rest.

122 See, D. Klepper and R. Kolinkow, Introduction to Mechanics, W.W. Norton & Company, Inc. (1968), starting at Chapter 3 on page 111 with special emphasis on pages 116-117 and See also, R. Weidner and R. Sells, Elementary Classical Physics, Allyn and Bacon, Inc. (1973), Section 8-4, pages 116-118 and Section 12-5, pages 221-223.

this center-of-mass will forever remain the same and whatever physical processes are occurring at that location keep a "perfect" time.[123]

Further, from the perspective of this unique, center-of-mass point, every subsequent coincidence of this point and a defined point in any other "uniformly[124]" moving reference frame would inevitably be a repetition of the famous "twin paradox" or "clock paradox" thought experiment and, since virtually every derivation of this experiment that the author has seen reaches the conclusion that the twin who was stationary will be older in absolute terms than the twin that was accelerated, and, moreover, that the experiences of the two twins will be different and that each will agree on the differences,[125] it follows naturally that all

---

123 In this regard, let the author suggest a problem for further study. Begin with a system that has three clocks all located immediately next to each other and all at the center-of-mass of a universal system. Place identical rockets on either side of the center-of-mass with a single immensely massive fuel tank between the rockets for both to use. Next, attach a cable from each rocket to just one clock so that each rocket will pull one clock away from the preferred center-of-mass location. If one instantly expended the fuel in the common tank and treated each rocket as the exhaust for the other, would the now slower readings on the then accelerating clocks suggest any change in the mass of the overall system? If one changed the situation so that the rockets were not identical and thus accelerated away from the center-of-mass of the system at different rates (although with the same momentum), what readings would the clocks give and how would these readings reflect on the mass of the overall system?

124 The point we are discussing is the center-of-mass and this center-of-mass also would be the center of gravity for the system. See, R. Weidner and R. Sells, Elementary Classical Physics, Allyn and Bacon, Inc. (1973), Section 12-5, pages 221-223. The author places "uniformly" in quotations, then, because every reference frame approaching the center-of-mass, from any direction, would be accelerating toward that center and thus could not be moving at a constant speed.

125 See, French, A.P., Special Relativity, W.W. Norton & Company, Inc. (1968), pages 154 to 159; and See W.G.V. Rosser, An Introduction to

will agree that the center-of-mass reference frame is inevitably "different" and has a perfect clock.[126]

Indeed, if one pays attention to the gravitational influence of the fuels that would be required to create all of the various accelerations that are part of the famous "twin paradox" experiment, one would be all the more certain that the two observers could distinguish their situations. To see that this is true, imagine an attempt to bring a reference clock within a universal system in which the conservation rules outlined above apply into coincidence with the system's center-of-mass two or more times. To accomplish this, one must take the following steps in sequence at least once and perhaps more than once:

- ◆ Stage 1: One must, by an expenditure of some fuel or performance of some work, accelerate the reference object that is or will be coincident with the center-of-mass in a direction that will lead it away from the center-of-mass, all the while paying attention to the continuing interaction between this object and the launch fuel (both expended -- i.e. the exhaust -- and unexpended);

---

the Theory of Relativity, Butterworth & Co. (Publishers) LTD. (1964) at sections 11.1 beginning on page 397 and 12.5 beginning on page 445.

126 As noted above, there is also authority for the proposition that all would agree on the location and velocity of this unique center-of-mass frame. See Rindler, W. Essential Relativity, at Section 5.7 (Springer-Verlag 1977), where the author appears to establish: (1) that there is a unique center of momentum frame of reference for any system of objects; (2) that this center of momentum frame is also the center-of-mass frame: (3) that the mass in the center-of-mass or center of momentum frame corresponds to the "rest" mass of the system if its composite nature were not recognized; and (4) that observers on all objects in the system would agree on the "mass" and "velocity" of the center of momentum frame. A consequence of this proposition is that any observer that moves with respect to this system would suffer measurement distortions (as described by the Lorentz Transformations) that prevent his identification of an absolute frame of reference based only on local experiences.

- Stage 2:   One must bring this object to a stop with the expenditure of some additional quantity of fuel or the performance of some additional quantity of work (although the gravitational pull of the gravitating Stage 1 fuel would assist in the deceleration process); and

- Stage 3:   One must accelerate this reference object back toward the center-of-mass by the expenditure of still more fuel or the performance of some additional quantity of work. In this instance, the gravitational pull of the Stage 1 fuel would assist in the re-acceleration process and the gravitational pull of the State 2 fuel would oppose the process)

Note, of course, that, in order to accomplish stage 3, it normally[127] would be necessary in stage 2 to decelerate a system that includes both the reference object and the fuel that would be necessary to re-accelerate this object to the speed with which it will return to the center-of-mass. Further, in order to accomplish stage 2, it would normally be necessary in stage 1 to accelerate an assembly that includes the fuel necessary to bring the reference object/re-acceleration fuel assembly to rest. In the absence of gravity, of course, each of these exercises would be possible, regardless of the ultimate speed at the second co-incidence with the center-of-mass, since the gravitational influence of the potential energy that must be stored for later use can be ignored.

In reality, however, the gravitational pull of the fuel to accomplish stage 3 would inevitably interact with the mass of the fuel to accomplish stage 2 which, in turn, would inevitably interact with the mass of the fuel required to accomplish stage 1. Since, as discussed above, there must be a maximum aggregate acceleration (and thus velocity) for an object of any finite mass that interacts with the fuel that is required to accelerate that mass,

---

127 [127]The author inserts the word "normally" here because it seems conceivable that the launch fuel might be so massive that its continuing gravitational interaction alone might be adequate to bring the object to a stop and return it to the center-of-mass at the right speed.

then, for each potential coincident object, there is a maximum speed of second coincidence. To allow a greater speed of second coincidence is to create a system that is so massive that the stage 1 assembly would collapse under its own gravitational pull. In a closing section of this work, the author places such an exercise in a specific context by approximating (without gravity) the launching of a 1000 kilogram mass at a speed of .999 percent of the speed of light, the deceleration of this mass to rest and the re-acceleration of this object to .999 percent of the speed of light back in the opposite direction in an effort to highlight how quickly the mass of such a system grows in the context of the required multi-step process[128].

Note also that, for the two observers to start in "identical" situations, one would need to locate both adjacent to identical quantities of fuel at the start of the experiment so that each would sense the same gravitational pull before the experiment began.[129] Returning, then, to the author's assertion that the two observers would not have to rely only on acceleration perceptions to determine which had done what, note that the observer who travels would both feel a substantial acceleration when his or her rocket fired and would also perceive the ever diminishing pull of gravity from the fuel that has been ejected from his rocket and the diminishing gravitational pull of the ever more distant unexpended fuel in his companion's rocket as well. The stationary observer would also be able to detect evolving gravitational "fields". This observer would feel the continuing presence of the fuel placed in his rocket and would also feel

---

128 The formal discussion of the photon rocket found in French, A.P., Special Relativity, W.W. Norton & Company, Inc. (1968), pages 183-184, provides further support for the idea presented. The example provided by Mr. French indicates that, assuming a speed of .995 percent of the speed of light, the payload of a photon rocket that is accelerated away from a point of origin, decelerated at a distant point, brought back to the origin and stopped again at the origin could only be 1/10000 of the total mass of the original rocket plus fuel assembly.

129 Even doing so would not entirely create identical initial experiences since the fuel in the rocket of the other traveler would create an asymmetrical initial "field".

the influence of his companion's fuel and his companion's rocket as well. The actual perception of the gravitational pull of the fuel and the companion rocket, of course, might be quite limited because our observer would be almost coincident with the center-of-mass of the system that consists of the traveling observer and this observer's exhaust.

f.   Closing Comments on the Theoretical Underpinnings of General Relativity

To complete the discussion on this topic, the author would note that Special Relativity was born when the 19th century idea of locating a preferred reference frame with electro-magnetic experiments came to naught. In hindsight, of course, it is clear that the location of a preferred reference frame through electro-magnetic means was doomed because of the conservation of net electric charge, the inverse-square nature of Coulomb's law and the invariance of Maxwell's equations when moving between different reference frames.[130]   Because of these features of the electro-magnetic interaction, it is simply impossible to locate a preferred frame by observing differences in what we measure as electro-magnetic experiences.

Pre-19th century physics, of course, had grown comfortable with the idea that it was impossible to locate a preferred reference frame in reliance on experiments in mechanics as well. This belief was intimately related to the then-existing concept of a mechanical "inertial" frame of reference. According to the Stanford Encyclopedia of Philosophy:[131]

> an inertial frame is a reference-frame with a time-scale, relative to which the motion of a body not subject to forces is always

---

130 See W.G.V. Rosser, An Introduction to the Theory of Relativity, Butterworth & Co. (Publishers) LTD. (1964) at section 7.6 beginning on page 300.

131 This is an "open access" resource available over the internet. The article quoted was entitled Space and Time: Inertial Frames and is listed as first published March 30, 2002 with a substantial revision on November 4, 2009.

rectilinear and uniform, accelerations are always proportional to and in the direction of applied forces, and applied forces are always met with equal and opposite reactions. It follows that, in an inertial frame, the center-of-mass of a system of bodies is always at rest or in uniform motion. It also follows that any other frame of reference moving uniformly relative to an inertial frame is also an inertial frame.

Although the concept that one can only detect relative motion rather than absolute motion through *mechanical* means is the absolute cornerstone of the theory of General Relativity, there is, to the author's knowledge, no reference in all of the literature of physics and certainly nothing in the literature of General Relativity to any single experiment or group of experiments that establishes the validity of this concept. Albert Einstein, of course, did not "prove" this concept -- he was a "theoretical" physicist after all -- and simply assumed its validity. The author has read that the "proof" of this idea -- the idea that all mechanical inertial frames are equivalent -- was provided by Christian Huygens. See W. Rindler, Essential Relativity, at Section 1.11 at foot note 4 (Springer-Verlag 1977). If this is the case, one can be sure that Mr. Huygens, a contemporary of Isaac Newton, did not use the precepts of Special Relativity in his proof. Indeed, the term "inertial frame of reference" was apparently coined by Ludwig Lange in 1885. See Wikipedia, the free encyclopedia, the article entitled "Inertial Frame of Reference." Here again, given the date the term was coined, we can be sure that the mechanics of Newton and not of Special Relativity were firmly in mind when the concept of the existence of interchangeable inertial mechanical reference frames with wildly different velocities finally crystalized.

In fact, if one reads the passage quoted above from the Stanford Encyclopedia of Philosophy carefully, it should be clear that the Newtonian concept of the interchangeability of each "inertial frame" required the invariance of the concepts of "force" and "acceleration" when moving between reference frames in uniform motion with respect to each other[132]. Acceleration is an invariant quantity under the Galilean transformations appropriate for application of Newtonian Mechanics. Significantly, neither force nor acceleration is invariant under the Lorentz Transformations. Both require application of extremely complex formulas to translate a force or acceleration as measured in one Lorentz frame into a force or acceleration as measured in another. See W.G.V. Rosser, An Introduction to the Theory of Relativity, Butterworth & Co. (Publishers) LTD. (1964) at section 5.5 beginning on page 202 with special emphasis on pages 206 to 207 for force and section 4.2 beginning on page 142 for acceleration. Significantly, the Lorentz Transformation formulas for both force and acceleration contain velocity dependent terms, and, moreover, force and acceleration are no longer co-linear. The author therefore challenges any adherent of General Relativity to step forward with a proof of the equivalence of all *mechanical* reference frames using Lorentz Transformation equations in the process[133]. Absent such a proof, the foundations of General Relativity rest on smoke and the elaborate system of mirrors used in the experiments of Michelson and Morley.

---

132 Many scientists now are critical of Newton for comments indicating his belief that both time and space are absolute. It should be remembered, however, that Newton's data set consisted of measurements, mostly by astronomers like Copernicus, in a single frame of reference, that of the Earth. Because of the link between measurements of space and time and momentum and energy as now evident in Special Relativity, there was no way for Newton not to reach his conclusion of the absolute nature of these measurements. It is only when we contemplate rapid motion with respect to the Earth that we can begin to see a reason that measurements from different frames of reference might suggest problems with the Newtonian approach. On the other hand, because of the links between space and time and momentum and energy evident in Special Relativity, we are forced back to a belief that there must be an absolute frame albeit a frame that we have yet to identify.

133 A comprehensive and rigorous proof of this fact using the Galilean transformations is provided in Elliott, R. Electromagnetics, at Section 2 of Chapter 2 starting on page 41 (McGraw-Hill Book Company, 1966).

Indeed, to the author, it seems impossible to have two forces with identical inverse-square force rules yet with one in which "charges" are conserved regardless of their state of motion and the other in which "charges" (i.e. rest masses) vary with motion and to nevertheless assume that there is not an absolute frame perceptible through measurements made with at least one of these forces. In fact, if it takes the requirements of conservation of charge and invariance of field equations coupled with an inverse-square force law to prevent the measurement of absolute motion by electro-magnetic experiences, isn't it certain that it is possible to measure absolute motion by gravitational experiences since gravitational "charge" definitely is not conserved and there are no precise analogs to Maxwell's field equations for gravity?

Moreover, if there is only so much gravitating mass and energy in the universe -- i.e., if the sum of mass and energy is conserved -- can we not inevitably find the frame of reference in which the quantity of mass and energy is minimized and declare this frame unique and co-extensive with the universe.[134] Every other frame is just a frame of con-

---

134 See again, Rindler, W. Essential Relativity, at Section 5.7 (Springer-Verlag 1977). The author must frankly add that he has always had difficulty with the concept that the frame of reference of a subatomic particle accelerated to tremendous speeds in a circular accelerating system on the surface of the earth (and thus still constantly accelerating even though the "speed" of the particle is unchanging) could be considered to be complete with observers able to report events to it based on an infinite coordinate system that stretches in all directions and with this frame equally valid for the description of all phenomena as the frame of the center-of-mass of the solar system. What is the shape of the sun in the frame of reference of this particle when the particle is directly approaching or directly moving away from the sun? What is the shape of the sun when the particle is moving at right angles to our line to the sun? What would be the mass of the energy implied by these rapid deformations and do we perceive the gravitational influence of this energy? Obviously we don't and, indeed, as some level, we believe with confidence that the solar center-of-mass frame of reference is a more "real" approximation of the universe than the subatomic particle's. We

venience, a localized area in which subsystems of the larger universe can be analyzed.

Further, given the variation of mass with velocity and, therefore, the inevitably escalating gravitational interaction between the required "equal and opposite" momentum carriers implied when viewing rapidly moving actors in a universe with no net momentum because its center of mass is at rest, isn't it clear that the historical tendency to attempt to analyze isolated systems in purportedly "inertial" frames grows less and less defensible as one deals with objects that are "observed" to be moving at greater and greater velocities. The fact that every object observed to be in motion implies another object or set of objects of equal but opposite momentum[135] mandates that the influ-

---

believe this in part because we know we have carefully orchestrated the local conditions of the universe to create the precisely calculated motion of this particle and that, in this effort, we have done nothing that would have a material impact on the sun given our understanding of physics. On another level, however, we believe that the solar center-of-mass frame is ultimately the "more" real -- although "less" real than a universal center-of-mass system -- because this returns us to a world view championed centuries ago by science against established religious dogma and returns us there for the same reason. The description of the universe from the solar center-of-mass view is radically simpler. It is, nevertheless, possible to describe the universe from the perspective of the rapidly moving and rapidly accelerating subatomic particle, just as it was possible (and still is possible) to describe the universe from the perspective of the moving and accelerating earth as was done by Ptolemy, his sources and his contemporaries. The same approach to science that led physics to abandon the earth as the center of the universe, then, argues against the generalization of the "relativity" principle in the way Einstein has done. If Einstein was and is right, on what basis could we reject the geocentric universal views of Ptolemy and the ancients?

135 The motion of one object or system implies the motion of another object or system because the center-of-mass of an entire system must either be at rest or move uniformly and the uniform motion of the center-of-

ence of the "off-screen" equal and opposite movers be considered if the "on-screen" actors that are under scrutiny have a sufficiently large mass in the frame in which they are being analyzed. There would be a gravitational link between the on-screen actors and off-screen equal and opposite movers that simply should not be ignored. This is not, of course, to say that these equal and opposite movers will inevitably be found. Instead, it is to say that, if these equal and opposite movers are not found, we need to look for the off-screen equal and opposite movers that are related to our observer to see whether it is the observer that is moving rapidly. To be sure, all manner of possibilities suggest themselves as to relative motion (i.e., both the observer and the actor may be moving) but this relative motion must be totally integrated into a universal view. An analysis of a sub-universal system is inevitably only an approximation and the approximation is less and less valid the more massive the system under analysis[136].

---

mass of the entire universe would establish a preferred direction which is presumed not to exist.

136 Some time ago, the author watched a television show with the title of "Mythbusters" and noted an episode in which the cast tested the "myth" of a Civil War incident in which two soldiers aimed their rifled muskets at each other, fired the muskets simultaneously and the bullets struck each other in mid-flight resulting in two lead miniballs fused into a single mass that fell near the point of impact. Because of problems in firing two such period weapons simultaneously, ultimately the cast abandoned attempts at a simultaneous firing and opted to discharge one weapon aimed at a stationary bullet. The result was the single mass alleged to have been produced in the "mythical" incident that was being tested. The relevance of this "experiment" to the present discussion is that, while it is possible to adopt a frame of reference in which the bullet fired from the gun is the moving object, it is also possible to adopt a frame of reference in which the bullet that we know to have been stationary is, in fact, the moving one. A center-of-mass frame in which the bullets approach each other at equal speeds is also possible. Indeed, to devout relativists, it is possible to adopt a whole host of hypothetical frames in which the relative velocities of the bullets vary over a broad range with one or the other almost infinite. Significantly, these relativists would further assert

In this regard, let the author yet again come back to Purcell's introductory text on Electricity and Magnetism referenced above and the discussion of the assembly of an electrically charged sphere that is found there. The Purcell discussion indicates that, if one were assigned the task of constructing an electron from a charged sphere of infinite radius and appropriate surface charge density, one would be required to perform work in a calculable quantity against the aggregate repulsion of all of the surface charge elements throughout each incremental volume of space. The total amount of work done would depend on the final radius of the sphere, to be sure, but the amount of work done to reach a particular radius and also the quantity of work done in each incremental volume located outside this radius of the sphere would be fixed quantities. The additional work done after a particular radius is reached has no impact on the work already done elsewhere.

Accordingly, the construction of a single stationary electron implies the creation of a unique universal space with each incremental volume of that space imbued with a defined quantity of something that we have come to call energy and can either be viewed as the work done against electrical repulsion within this incremental volume of space in the construction of this electron or, simply, as the energy cost of creating the electric field within this volume[137]. This, to be sure, is not an original

that all of these frames are of equal validity. We know this to be false, however, because the universe contains the recoiling gun from which the moving bullet was fired and the expanding gasses from its muzzle and no change of reference frame can make these objects disappear or change the gravitational links between them, the bullet fired from them and the bullet that was struck.

137 The author has trouble understanding how it would be possible to alter this uniqueness and thereby make applicable Einstein's relativity principal if all we can do is install other similar unique field-producing objects -- electrons and protons and the like -- and superimpose their fields. The uniqueness of each volume of the field of a charged particle should insure that a universe consisting of charged particles and nothing else is similarly unique. Given that there are no uncharged stable

idea -- the author has read a number of different analyses that suggest that, if an electron has a radius in the range of roughly 10^{-13} cm, then the inertial "mass" (and thus gravitational mass based on our assumption that inertial and gravitational mass for discrete objects are identical) of a stationary electron is simply equal to the energy implied by its construction from an infinitely separated dilute charge density. See, e.g., M. Born, Einstein's Theory of Relativity, Dover Publications, Inc. (1965) at 207-214.

Significantly, late 19th Century and early 20th Century physics was poised to expand on this concept to start to unify electro-magnetic and inertial phenomena and had already formulated a mathematical basis for the increasing inertial mass of moving charged objects. See again, M. Born, Einstein's Theory of Relativity, Dover Publications, Inc. (1965) at 212-214. See also, A. Miller, Albert Einstein's Special Theory of Relativity Emergence (1905) and Early Interpretation (1905-1911), Addison-Wesley Publishing Company, Inc. (1981). The author believes that these older ideas have new relevance and they form a significant portion of his conception of a proper understanding of gravitation as detailed later in this work.

The unification of electro-magnetic and inertial phenomena suggested by these century-old papers is obviously only part of the story since the mass of a proton cannot be attributable to the historical accounting for the work done in bringing the positive charge elements in to create an electron-like object. The end result of such an exercise would simply be a positron and would have only the electron-equivalent mass of a positron. Nevertheless, a proton under current thought has a substructure that interacts through a force other than the electromagnetic force and it might be found that the added mass of a proton represents the historical accounting for the work done against this additional

---

particles with a rest mass, as noted above, it follows that the universe is the superposition of the fields of particles with unique impacts at every spacial point and, thus, that every spacial point is unique.

force in assembling the components of the proton's substructure from spherical proto-structures with infinite radii (or from a compact point source if the force is attractive)[138]. In this regard, the author would note that nucleon numbers are generally conserved much as electric charge is conserved so that the location of an absolute frame by reference to nuclear interactions might well be impossible just as it is for electro-magnetic interactions. It may also be that the dimensions of nucleon structures, which are far less than the dimensions of an electron, re-quire much greater energy -- the energy required to create the field within the difference in the electron radius and the nucleon radius such that the mass of a proton is also entirely electrical in origin[139].

---

138 The fact that an electron and a proton precisely neutralize each other means that the work constructing a proton from an infinitely diffuse charge density must start with the exact same density at infinity as an electron. Otherwise, there would be a residual energy function outside the radius of the electron and the divergence of this energy function would be a residual electric field.

139 The author feels that the now experimentally well documented concept that energy has a mass equivalence and an expectation that there are purely "attractive" forces between mass elements and objects are deeply incompatible. The reasoning is simple. As two attracting objects are separated, work is done against the attractive force and the space through which this work was done acquires its own "mass." How, then, can the interaction of the attracting objects and this new mass element not result in a feedback spiral that insures collapse. It is the contemplation of this problem that led the author to suggest the complementary approach to electricity and gravitation that is outlined in footnote 21, above. A universal system that includes a primary repulsive force -- even of two varieties with each variety attracting its opposite but repelling its like -- but prevents this primary repulsive force from ever destroying any of the "energy" content that it stores at any one moment of "time" as defined in one master frame of reference (i.e., the center-of-mass/energy frame) seems a necessary approach to our current universe. The energy constraint affects the behavior of the primary repulsive field in such a way that it appears to create non-electrical "forces" that influence the boundaries of field sources and therefore causes the appearance of "movement" by the boundaries of these field

Such an approach -- an approach in which mass is an attribute that affects local areas of space based on electromagnetic occurrences in that space over time -- would deal with the problem identified in a footnote above regarding a "finite" velocity for gravitational interactions. As discussed in that footnote, such a finite interaction velocity suggests the creation of a perpetual motion machine. To repeat in part the presentation above, imagine allowing an electron/positron pair to annihilate each other and generate two gamma rays moving in opposite directions. It appears that these photons could not interact gravitationally under current thought because their relative velocities exceed the maximum speed of the interaction and thus neither can sense the other. Our "machine" would capture these photons after they had separated and the objects that captured these photons would be allowed to collapse into each other. The mass enhancement associated with the captured gamma rays would interact gravitationally in the process of collapse and thus would allow work to be done on the return stroke that was not done in the separation process. In contrast, if we assume that the historical work done in a finite and specific distant volume of space in the creation of the two particles is unchanged after the assembly, then the creation, separation and annihilation -- all done consistent with Maxwell's equations and giving due consideration to the energy and momentum resident in electro-magnetic fields[140] -- would do nothing to this distant space except as the induced electromagnetic consequences of the events are propagated to the relevant distant space.

As a further commentary on the theoretical underpinnings of relativity, the author would suggest that the steps leading to the "discovery" of

---

sources and thus of the sources as well. The "motion" required by the energy constraint obeys a force law that mirrors the force law of the primary force just as the shadow of an object obeys laws of motion that are inextricably linked to the motion of the object itself.

140 See Tolman, Richard C., Relativity Thermodynamics and Cosmology, Oxford at the Claredon Press (1958), Section 42, page 89-90.

the relativity principle represent almost as much of a historical anomaly or artifact as the Newtonian "assignment" of infinity as the zero-point for gravitating systems. Early physicists, applying the rules of Newtonian Mechanics -- and, significantly, ignoring gravity -- concluded that they could not locate a precise single universal frame of reference. Based on the mechanics with which they were familiar, they could always envision the acceleration of ever more immense objects to ever more incredible speeds simply by ejecting a trivial object -- a peppercorn, to borrow a legal term – with sufficient velocity and therefore energy. To these physicists, energy did not interact with massive objects and so there was no such thing as a system that was too massive to be escaped and no speed limit. According to these pioneers, if there is no speed limit and no interaction between massive objects and the energy that moves them, all speeds are possible for all objects. Further, to these physicists, it was always possible to dilute the force of gravity with distance. Conceptually, these ideas plus the invariance of acceleration between Newtonian reference frames translated into a supposition that each point in space could be part of an infinite number of "inertial' reference frames since an observer -- even an observer tied to an object of incredible mass -- at this point could envision additional objects of almost infinite mass moving at almost infinite speeds in any direction and off to infinity[141].

On the other hand, were these early physicists pressed, they likely would have expressed a belief that they simply lacked the tools to locate an absolute frame and not that there was no such a frame -- given the strength of organized religion at the time, these physicists likely believed that the frame of reference in which their deity operated was the

---

141 As noted above, a comprehensive and rigorous proof of this fact using the Galilean transformations is provided in Elliott, R. <u>Electromagnetics</u>, at Section 2 of Chapter 2 starting on page 41 (McGraw-Hill Book Company, 1966). <u>See also</u>, T. Frankel, <u>Gravitational Curvature (An Introduction to Einstein's Theory)</u>, W. H. Freeman and Company (1979) at 1-2. <u>See also</u>, the Discussion at 9.e., above.

ultimate frame of reference of the universe. Indeed, the experiments that ushered in the mechanics of Special Relativity (the Michaelson/ Morley experiment, for example) were performed by physicists confidently searching for the ultimate rest frame -- the rest frame of the ether[142]. Later physicists, of course, turned the lack of confidence in an ability to locate a universal frame into an affirmative assertion that no such frame exists. Einstein (and, of course, many others), therefore, naturally took experiments that established that no amount of analysis of *electro-dynamics* could identify a unique reference frame as confir-

---

142 Modern consensus, of course, has done its best to banish the electromagnetic or lumiferous ether from current thought and to relegate it to history. A careful reading of various contemporary analyses of electrodynamic phenomena, however, leads inescapably to the conclusion that there is, indeed, a universal "medium" that pervades all of space. The ether is the electro-magnetic field itself -- or, more properly, the superposition of all of the electro-static fields of all of the charged "particles" in the universe but with each electro-static field converted to the appropriate "electric" and "magnetic" field components appropriate to the relative state of motion of the nominal "source" "particles" as adjusted to the frame of reference of the observer in search of his or her "ether" and also with due consideration of the energy content resident in the electro-magnetic field based on the entire history of the motions of all field sources. Ancient prejudices lead us to focus on the "holes" in this universal ether -- the places where the field is not, such as the interior of an electron or proton and the boundaries between the ether and its holes -- rather than on the remainder of the universe where electric fields are busy inducing magnetic fields and magnetic fields inducing their electrical counterparts in an intricate series of interrelated steps ultimately dictated by considerations of mass/energy conservation, the constraints of Maxwell's equations, the requirement that charge be conserved and the requirement that charges be quantized. The ultimate current consensus that the laws of *electrodynamics* prevent the local detection of absolute motion does nothing to establish that there is no preferred frame ascertainable by a detailed analysis of *mechanics*. Indeed, the age old reference of motion to the "fixed" stars and the ancient sciences of navigation and geography make it clear that there is, indeed, a substantially fixed framework against which motion can be measured.

mation of the previous proof of the absence of a fundamental *mechanical* reference frame.

These physicists ignored the constant criticism of the opponents of Special Relativity that its two fundamental tenets were irreconcilable. Opponents of Special Relativity, most now long dead, properly argued that the mechanics they understood had to produce a unique frame that was detectable through experiments with electricity and magnetism. Proponents naturally responded with the obvious -- electromagnetic experiments of myriad design and execution had repeatedly and consistently failed to detect such a frame. Today, only proponents seem to survive and they certainly can point to an impressive list of experiments that support the proposition that no analysis of electrical or magnetic fields allows the discovery of an absolute reference frame. Further, the success of the mechanics of Special Relativity, which was built on the premise that electro-magnetic phenomena must have consistent measurements throughout all potential reference frames, now stands as independent proof of this fact.

On the other hand, if one starts with two logical propositions (say A and B) that cannot both be true; if one develops experiments under the assumption that A is true to establish that B is false; if the results of those experiments fail to prove that B is false and one thus comes to conclude that B is true, one cannot then conclude that *both* A and B are true. If one is confident that B is true, then A must be false.

Assuming that both of these irreconcilable propositions were true, however, is precisely what Einstein did[143]. Einstein left undisturbed the pre-existing assumption of his adversaries that there can be no

---

143 See A. Einstein, <u>On the Electrodynamics of Moving Bodies</u>, Introduction, *reprinted in* <u>The Principle of Relativity</u>, Dover Publications, Inc.(1952) at pages 37 to 38 where Albert Einstein himself (or more properly, the translator of his German text to English) said (with emphasis added by the author):

absolute significance to *mechanical* frames of reference and developed a completely new mechanics based on the assumption that *electrodynamic* measurements also cannot establish anything other than relative motion.

To the knowledge of the author, however, no one has ever attempted to go back and prove, using the new mechanics of Special Relativity, that all observers reach determinations of mechanical events that are of equal validity. Instead, every work the author has read -- and the author has read (or at least skimmed) quite a few -- simply establishes the absence of a universal reference frame discoverable through Newtonian Mechanics (without gravitation), then references the ether drift experiments and related experiments in *electro-dynamics* as proof that there are no detectable preferred reference frames as expected by then existing mechanics and then concludes that there can be no preferred reference frames detectable by *any* means either *electro-dynamic or mechanical*. Although these works all continue and develop

---

Examples of this [disparate analyses of electrodynamic phenomena depending on perspectives having a moving conductor and stationary magnet or a moving magnet and a stationary conductor], together with the unsuccessful attempts to discover any motion of the earth relatively to the "light medium" suggest that the phenomena of *electrodynamics* as well as of *mechanics* possess no properties corresponding to the idea of absolute rest. They suggest rather that, as has already been shown to the first order of small quantities, the same laws of electrodynamics and optics will be valid for all frames of reference for which the equations of mechanics hold good. We will raise this conjecture (the purport of which will hereafter be called the "Principle of Relativity") to the status of a postulate, and also introduce another postulate, which is only apparently irreconcilable with the former, namely, that light is always propagated in empty space with a definite velocity *c* which is independent of the state of motion of the emitting body.

See also, B. G. Bergmann, Introduction to the Theory of Relativity, Prentice Hall, Inc. (New York, 1947) at 28 (and note again that Albert Einstein wrote the forward to this book).

the new mechanics dictated by Special Relativity, they never return to the point of departure to confirm the original supposition made with the older faulty mechanics that there are no absolute *mechanical* reference frames.

If they did return and attempt to prove this point, the author believes they would encounter problems that cannot be solved[144]. To repeat an example suggested previously, consider two objects of widely different rest masses that are both near a third substantially larger gravitating mass. Assume that the two objects at rest are set in motion by a single event, that, in this motion, the two objects begin to move at right angles to the line back to the center of the larger gravitating mass, and that at least one object is now moving at a relativistic speed as compared to the speed of the previously existing assembly at rest. The speed of the more massive of the objects, of course, would be less than that of the less massive object and might be many orders of magnitude less[145].

When an observer moving with the faster of the two objects sits down to compute the dimensions and mass of the large gravitating mass using the new mechanics of Special Relativity, that observer will see the dimensions of this large mass foreshortened in the direction of the faster object's motion by the familiar Lorentz contraction factor of $1/(1-v_{faster}^2/c^2)^{\frac{1}{2}}$ and it will see the mass of the larger gravitating mass enhanced by the factor of $m_{Large\ Body}/(1-v_{faster}^2/c^2)^{\frac{1}{2}}$. The slower of the two now moving objects will, again using the mechanics of Special Relativity, see the dimensions of the large gravitating mass foreshortened in the direction of its own motion but only by a factor of $1/(1-v_{slower}^2/$

---

144 The problems would become all the more significant if they fully incorporated "gravity" as inevitable feature of mechanical systems that must be considered in their analysis but we will not attempt this exercise yet.

145 Imagine a space ship that has been designed to use a device similar to a particle accelerator to eject a stream of protons as its exhaust and that the protons are accelerated to say .999 percent of the speed of light before they are ejected.

$c^2)^{\frac{1}{2}}$ and will also see the mass of this larger object enhanced but again only by the factor of $m_{Large\ Body}/(1-v_{slower}^2/c^2)^{\frac{1}{2}}$. In contrast, an observer on the large gravitating mass will see the now moving objects both fore-shortened and of enhanced mass although the change in dimensions and mass enhancement of each object will depend on each relevant object's speed[146]. The mass of the large gravitating object will not be changed as far as this observer is concerned.

Accordingly, neither of the now moving observers will be able to make perfect sense of their gravitational experiences based on their individual computation of the inertial mass of the large gravitating object. At the instant of acceleration, however, each object will be in an identical gravitational "field" since their motions are at right angles to a direct line between their original position and the center of the large gravitating mass. In the final analysis, the conclusion is inescapable that the inertial masses of the two now moving objects are enhanced in the universal frame of reference as a result of *their* motion[147]. Further,

---

146 Note also that mass enhancement of each moving object will have to have been obtained by the conversion of something other than the mass of either object into kinetic energy. Thus, in truth, there would have had to have been 4 system components originally -- the now rapidly moving object, the slower moving object, the large gravitating mass and the tank of fuel that was converted to the kinetic energy of the former two objects.

147 To confirm this, the author produces the following from W.T. Scott, The Physics of Electricity and Magnetism, John Wiley & Sons, Inc. (1966) at section 10.3, beginning on page 543 with the quoted language taken from page 553:

> The Lorentz-Fitzgerald contraction can thus be described as being an expression of the foreshortening of the electric filed of a moving charge. It is not surprising, therefore, that a macroscopic piece of material composed of electric charges will also foreshorten. It was the contribution of Einstein to show that this contraction is general and not just a property of the electric fields of the charged particles. Although his account of the contraction depends on the nature of simultaneous observations of two points on a moving body and

these two observers could not agree with each other as to those inertial masses because their velocities cannot be the same. On the
other hand, every other potential observer who viewed the original two
objects at rest and watched the event that caused their motion will see
our two original observers as those that are in motion because each's
motion causes the other's and all will agree on this and all will agree
that energy and momentum and angular momentum, etc., were conserved in the process by which the motion began. Both General and
Special Relativity, however, assume that all three of the observers we
have discussed -- the one moving at a relativistic velocity, the one moving much more slowly and the one on the third still larger "stationary"
mass -- can claim that their analysis of the universe is as valid as the

---

thus appears to be quite different than the treatment given here,
the fact that a common observation time *t* is used here and that the
fields may be found from a Lorentz transformation of the Coulomb
field shows that the two derivations are actually closely related.

The author's view is that the charged components of the moving
object are "foreshortened" as an expression of the "foreshortening" of
the electric field of a moving charge and that this is general because
the object consists only of the moving charges and the (stabilizing)
arrangement of electromagnetic fields that bind the moving charges
together. Indeed, in 1902, a similar discussion of the mechanism
and reality of the "foreshortening" was prepared by the very Lorentz
whose name is associated with the contraction. See H.A. Lorentz,
Electromagnetic Phenomena in a System Moving with any Velocity
Less than that of Light, at §8, *reprinted in* The Principle of Relativity,
Dover Publications, Inc.(1952) at pages 21 to 23. See also, M. Born,
Einstein's Theory of Relativity, Dover Publications, Inc. (1965) at 221.
Of course, we don't know which object is "really" "foreshortened" until
we have identified the master frame of reference. Perhaps, from the
point of view of this master frame, the object moving rapidly with respect
to our large object has really slowed and thus is less "foreshortened"
than it used to be. All we know is that there are changes in relative
dimensions of moving objects dictated by the rules of electrodynamics
and that the electric and magnetic aspects of those electrodynamic rules
give no local insights as to absolute motion.

other two. The equivalence of the viewpoints of these three observers, therefore, cannot be defended.

Indeed, with the fullness of the development of Special Relativity and science in general that has occurred in the now more than a century since Einstein's original work, it should be clear just how wrong the "relativity principle"[148] is. Relativity is a theory predicated on the ideal that the universe is filled with electrically neutral objects in various states of relative motion and that measurements of this motion can be made using electrically neutral measuring sticks and electrically neutral timing systems. Separate and apart from these neutral things, Relativity develops rules for the behavior of charged objects and their fields which are measured by reference to these various electrically neutral things. Sitting now, in the 21st century, however, we have a firmly-rooted atomic theory of matter. Consistent with such an atomic approach to the universe, we should realize that every observer and every apparatus of observation that has ever been or could ever be is built of electrons, protons, neutrons and the like which, in turn, are organized into atoms which, in turn, are organized into further aggregations such as molecules, crystals, etc. Every observer and every

---

148 A frequent description of the relativity principal is that the laws of physics are the same in all inertial reference frames. This formulation seems innocuous enough that it may seem self-evident. It is not. As an analogy consider that the rules of baseball are the same in all baseball parks -- the four base paths are always 90 feet long and the pitching mound is always 60 feet, 6 inches from home plate. Notwithstanding the identity of these rules in all parks, a game of baseball in Yankee Stadium is a different and unique experience as distinguished from a game of base ball in Fenway Park. In Yankee Stadium, the opponent would be the New York Yankees, a unique collection of players with unique talents. In Fenway Park, of course, the opponent would be the Boston Red Sox, a different collection of unique individuals. The same laws of electrodynamics govern play in both places yet the very fact that those laws provide for fields that are infinite in extent mean that every source charge in the universe has a unique and different voice in defining the local electrodymanics in Yankee Stadium and in Fenway Park.

instrument, therefore, is a participant in a grand universe of complex electrical and magnetic fields. There are no detached observers, and, further the "mechanical" behaviors of macroscopic "neutral" objects we perceive ultimately are based, on the microscopic level, on electrodynamics and nothing else.

The electrodynamics contemplated by Special Relativity, of course, tells us that all observers will come to related but distinct assessments of electrodynamic events. The electric and magnetic fields that make up the universe appear different to each observer and each apparatus because each observer and each apparatus is nothing more than the superposition of the electromagnetic fields of its constituent sub-atomic parts. Accordingly, if we set a single electron located on the earth in motion in a linear accelerator, that electron's assessment of the earth's, the solar system's and the universe's myriad fields will be radically different than its assessment when it was originally at rest. All of the other electrons, protons and neutrons in the universe, however, will continue to have the assessment of the universe's fields that they originally had -- except, perhaps for the additional assemblage of electrons and protons that must have been set in motion in order that momentum could be conserved as our electron took flight. As noted previously, then, it is not defensible to say that the frame of reference of this electron is of equal validity with all of the other reference frames available because there is no series of coordinated clocks located throughout the universe that are in sync with the clock of this single special electron and, of course, there are no rigid rods for measuring in the frame of this lone electron either.

Further, a variety of texts indicate that the sudden acceleration of this single electron would produce unique effects on the electromagnetic fields in the universe that leave no question whether it was the electron and not the rest of the universe that suddenly accelerated. As the accelerated electron changes from a state of rest to a state of motion, the change would generate an electro-magnetic wave centered on the formerly stationary electron[149]. This wave conveys both energy and momentum which are universally conserved quantities. The core principle of Special Relativity indicates that the field and the electromagnetic wave will have consistent (although different) features in all frames of reference, regardless of their state of relative motion. The idea, then, that this electron's calculation of the masses of all of the various existing components of the universe and its assessments of the distances between each of these components are on the same footing as the measurements of other observers -- take the measurements based on the earth's center-of-mass, for example -- cannot be right[150]. This notwithstanding, however, we certainly can (and, in fact, must) use the Lorentz transformation equations as applied to measurements made in this electron's frame of reference to calculate what the measurements of the different electric and magnetic field components of every charged particle that is part of the earth should be as measured in the earth's center-of-mass frame. It is only when we calculate the energy density of these field components and examine the universe for the electromagnetic waves that originate when a charged object accelerates that we could guess that one frame was superior to the other.

---

149 See E. Purcell, Electricity and Magnetism, (McGraw-Hill Book Company 1965) ¶5.7 beginning at page 163.

150 If we have a universe that consists of 100 equally charged objects and we set two of these in motion in equal and opposite directions, the two now moving objects will see changes in the fields of the other 98 while the 98 objects will see changes in only the fields of two. Not all of these observers can construct comprehensive and consistent frames of reference.

10. Comments on the Available "Proof" of the Validity of General Relativity

Notwithstanding the author's belief that General Relativity is a deeply flawed system for analyzing gravitating systems, the present scientific community believes with a surprising confidence that General Relativity's analysis is authoritative and that experimental verification for the theory is both substantial and convincing[151]. In contrast, the author accepts as authoritative only the experimental verification of the concepts of mechanics that are central to Einstein's Special Relativity theory. To the author, the localized verification of those concepts does not provide independent support for their generalization as is done in the development of General Relativity. Thus, while the author accepts Special Relativity as a thoroughly documented and

---

151 One dissident voice is that of Louis Brillouin whose work the author has referenced several times before. See L Brillouin, Relativity Reexamined, Academic Press (1970), Chapter 3, especially at Section 8, on pages 53-55; and Chapter 8, especially at Section 2., on pages 98 to 99. The discussion from Chapter 3 is telling:

> As a conclusion: There is no experimental check to support the very heavy mathematical structure of Einstein. All we find is another heavy structure of purely mathematical extensions, complements, or modifications without any more experimental evidence. To put it candidly, science fiction about cosmology -- very interesting but hypothetical.

> Altogether, we have no proof of the need for a curved universe (space plus time) and the physical meaning of the theory is very confusing.

experimentally well proven revolution in the understanding of the physical world, the author finds the alleged independent alleged verifications of General Relativity few and entirely flawed[152].

---

152 This, to be sure, is a rather bold statement if the literature on the subject is to be believed. There are some sources that suggest that General Relativity has been validated by experiments that confirm the predictions of the theory and which have left only a margin for error of less than two tenths of one percent. See, G. Tauber, Albert Einstein's Theory of General Relativity, Crown Publishers, Inc. (New York, 1979), Part III which begins on page 114 with special reference to the discussions on pages 132 and 145. With all deference to these sources, however, the author cannot accept the validity of any theory which has left undisturbed the erroneous zero-point convention that is discussed in the opening sections of this work and that also has failed to properly address gravitational potential energy as further discussed shortly thereafter. Indeed, a theory that has sent physics on a fruitless search of more than 20 years for the "other" 97 percent of the universe beyond what we can see and touch should already be regarded with suspicion rather than reverence. Accordingly, even if the author's discussion of the relevant data may be wanting in some respect, those with the time and sophistication to follow the many complications in the experiments that have been done should be able to find the errors now that it is clear what must be looked for. The author would hasten to add that he is not alone in finding the available "confirmation" of the principles of General Relativity to be inadequate to support the great weight of that theory. Again, according to Brillouin:

> Einstein introduced the adjective "restricted'" because he later tried to extend the principle [of relativity] to more general situations, but this extension was recently criticized in different countries by independent scientists who found many weak points in Einstein's assumptions. Many things happened since Einstein worked out his theory at the beginning of the century. Quantum theories invaded all chapters of physics, including mechanics and optics. Some of Einstein's assumptions looked safe, but they are now open to discussion and must be reexamined very carefully. While quantum theory helped us to discover many new phenomena in physics, we still have very few experimental checks of general relativity; it is time to go back to the "brave old relativity"

a.   An Inventory of Phenomena that have been Cited as Proof of General Relativity

These few independent alleged proofs[153], to the knowledge of the author, are based on three classes of phenomena[154]: (a) the precession of the perihelion of the orbits of the planets, especially Mercury[155]; (b) the deflection of electromagnetic radiation by gravitating bodies[156]; and (c) the "red" shift of electromagnetic radiation as it moves from points of lower gravitational potential energy to points of higher gravitational

---

and revisit all its territory. Every physicist feels that the very few (altogether three) experimental checks are really a meager result for too much computation. General Relativity is a splendid piece of mathematics built on quicksand and leading to more and more mathematics about cosmology (a typical science-fiction process).

See L Brillouin, Relativity Reexamined, Academic Press (1970), Chapter 1, where the quotation is taken from page 10.

153 Of course, in addition to these proofs are "proof" of the validity of the principle of equivalence. See, G. Tauber, Albert Einstein's Theory of General Relativity, Crown Publishers, Inc. (New York, 1979), Part III which begins on page 114. We have discussed problems with this proof in a prior section.

154 See, G. Tauber, Albert Einstein's Theory of General Relativity, Crown Publishers, Inc. (New York, 1979), Part III which begins on page 114. See also, L. Brown, A Pais and B. Pippard, Twentieth Century Physics, Institute of Physics Publishing and American Institute of Physics Press (Bristol, Philadelphia and New York, 1995) at Section 4.3.11 beginning on page 303.

155 See, G. Tauber, Albert Einstein's Theory of General Relativity, Crown Publishers, Inc. (New York, 1979), within Part III beginning on page 119 "Motion of the Perihelion of Mercury" and continuing at page 126 "The Relativity Effect in Planetary Motions".

156 See, G. Tauber, Albert Einstein's Theory of General Relativity, Crown Publishers, Inc. (New York, 1979), within Part III beginning on page 121 "Deflection of Light by a Gravitational Field" and continuing at page 125 "Gravitational Deflection of Light, Solar Eclipse of 30 June, 1973" and continuing again at 139 "The Time-Delay Test of General Relativity".

potential energy -- the lengthening of the wavelength of electromagnetic radiation as it moves away from a gravitating body.[157] A critic of General Relativity like the author, then, must do one of the following to undercut these alleged proofs: (1) challenge the existence of the relevant phenomena; (2) challenge the extent to which a phenomena has been accurately measured; or (3) show that correct measurement and understanding of the phenomena is inconsistent with the predictions of General Relativity. A critic to be convincing, of course, must also propose a mechanism by which these phenomena would occur and have their measured values but be the result of something other than space-time curvature.

b.   Comments on Gravitational "Red Shifts" as Proof of General Relativity

We will ultimately discuss each of these phenomena and the experiments that suggest their existence and significance in turn but let us start with the third -- the existence and extent of gravitational red shifts. The author will begin by stating his conviction that gravity does cause a "red" shift in radiation and that at least some of the experiments done to accurately measure that shift seem authoritative. Where the author parts company with the supporters of General Relativity is with the idea that such a shift is the result of "space-time" "curvature". Indeed, the author believes that the experiments which establish and quantify a "red" shift of radiation as it moves away from a gravitating body cannot be reconciled with the central tenant of the relativity theories -- both Special and General -- that there are no preferred reference frames. We will discuss why after outlining some of the experimental bases for concluding that gravity does result in a wavelength increase or

---

[157] See, G. Tauber, Albert Einstein's Theory of General Relativity, Crown Publishers, Inc. (New York, 1979), within Part III beginning on page 122 "Displacement of Spectal Lines Toward the Red" and continuing at page 132 "Terrestrial Measurement of the Gravitational Red Shift.

reduction for electromagnetic radiation as radiation moves away from or toward a gravitating body.

The first experiments to consider in this regard rely on the Mossbauer Effect[158]. The Mossbauer Effect is a resonance phenomena dependent on special radiation emission and absorption behavior of certain very special materials. Radiation emission frequencies of typical materials are offset from the absorption frequencies of the same substance because the emission and absorption typically happen at the atomic or molecular level. When such small structures emit radiation, the dual requirements of conservation of energy and momentum insure that some of the energy of the emitting excited atom or molecule is required to provide kinetic energy to the recoiling atom or molecule if both energy and momentum are to be conserved in the emission event. Similarly, an atom or molecule must be excited by radiation with sufficient energy to provide both post-absorption momentum as well as the excitation energy for the atom or molecule. Thus, emission results in radiation below the resonate frequency of a stationary atom and absorption requires radiation higher than the resonate frequency of that stationary atom.

Of course, the atoms or molecules that are emitting radiation typically will have a variety of relative thermal motions. Accordingly, even if each atom or molecule in a sample were to emit radiation at a very precise wavelength in its own rest frame of reference, the perceived wavelength of radiation from a sample spans the range between the emission wavelength plus the blue shift of the sample components moving toward the detector at the greatest of speeds available at a given temperature (adjusted, of course, for emission recoil) and the emission wavelength minus the red shift of the sample components moving away from the detector at the greatest of speeds at a given

---

158 The discussion of the Mossbauer Effect here is largely drawn from a much more rigorous analysis in W.G.V. Rosser, An Introduction to the Theory of Relativity, Butterworth & Co. (Publishers) LTD. (1964) in Appendix 6 beginning at page 493.

temperature (again adjusted for emission recoil). Indeed, because the detector is also typically a sample of material with component atoms in thermal motion, a detector atom or molecule will absorb radiation both above and below its precise rest absorption frequency. All these complications insure that very little of a typical emitting source's radiation can be absorbed by a stationary sample of the same material.

Materials that exhibit the Mossbauer Effect do not behave in this typical manner. In such materials, some emissions and absorptions apparently are not by discrete atoms or molecules but, rather, by a material fraction of the entire stationary sample. As a result, the range of frequencies both emitted and absorbed are unusually sharp -- there are neither material thermal variations nor material recoil complications. A substance with such sharp emission and absorption frequency lines is ideal for testing the existence of a gravitational "red" shift because, if gravitation causes such a shift, the radiation emitted by one sample at a certain distance from a gravitating body will not be absorbed by stationary samples of the same substance at greater or lesser distances from the same gravitating body. If, however, either the macroscopic source or detector is moved at the proper velocity to cause a compensating red or blue shift, then the emission and absorption lines will again coincide. The red or blue shift caused by the motion and the red or blue shift caused by the change in gravitational potential as a result of movement away or toward the gravitating body compensate for each other. The familiar formula for determining the "red" or "blue" shift caused by motion of one sample away from or toward an observer allows the gravitational shift to be quantified -- the gravitational shift and shift due to motion are equal at the point of greatest absorption and the gravitational shift can therefore be matched to the change in strength of the gravitational "field" between the two relevant points. Experiments have been done in reliance on the Mossbauer Effect that show that there is a measurable "red" shift of radiation of sources at a point of lower gravitational potential as compared to detectors at a higher potential (i.e. raised above the surface of the earth on a tower atop the source).

Classical gravitational theories, of course, ascribed no mass equivalence to radiant energy and, accordingly, radiation would not be affected by gravity. Experiments relying on the Mossbauer Effect, therefore, leave little doubt that Newton's classical theory of gravitation could not be the whole story. Then again, once Special Relativity required radiant energy to have a mass equivalence, there was no way to maintain the conservation laws that are so basic to physics without some gravitational interaction between massive bodies and radiation. Otherwise, as suggested at several points above, one could transmit energy away from a gravitating body in the form of radiation, suffer no energy loss in the process, allow capture of this radiant energy by an object at a point of higher gravitational potential energy, allow this object and its enhanced mass due to the capture of radiation to fall in the gravitational "field" and end, at the conclusion of a complete cycle, with enough energy for the object to rebound out to its old position, generate a new photon and have a little left over in the form of the kinetic energy acquired by the radiation-based mass enhancement of the original object during the course of the fall.

The gravitational red shift established by experiments relying on the Mossbauer Effect, then, in and of itself, is not proof of General Relativity, but, instead, is only proof that radiant energy and gravitating objects do interact. Standing on their own, these experiments merely establish that electromagnetic radiation loses energy and, therefore, the wavelength of this radiation is shifted to the "red" as that radiant energy moves "upward" and away from a gravitating object. Of course, relying on the traditional approach to gravitational potential energy (i.e. that such energy is zero at infinite separation), then the greater the gravitational potential energy -- the smaller the negative number that reflects that potential energy figure -- the further to the red the frequency of that radiation is shifted.

The most obvious problem with this experimental result is that it inevitably breaks the link between the presence of gravitating matter (and a gravitational "field" for that matter) and the slowing of clocks. To see that this is true, consider the situation in which a perfect clock is located at the very center of a spherically symmetrical mass distribution. The mass distribution that we will select in the first instance is just the familiar spherical shell of matter we have discussed previously. Let us assume that we can create holes with a very small diameter at the opposite poles of this shell. Because classical gravitational theory suggests (and General Relativity does not appear to suggest otherwise) that there is no gravitational "field" at the very center of such a shell (and our small holes can hardly have much impact on this)[159], a clock at the very center of the shell would be in a "field" free area (ignoring the "fields" of other objects in the universe as is the current custom). If an observer at this point sent signals outward to the surface and just beyond to observers above each of the poles, however, those signals would be shifted to the red because of the gravitational interaction of the radiation with the shell. The radiation, would, of course, be shifted further to the red as it left the surface and ultimately headed toward "infinity" -- gravitational potential depends linearly on distance and there are no places in a single object's traditional gravitational "field" where radiation would begin to shift back toward the "blue."[160]

We can only conclude, then, that it is not the gravitational "field" or gravitating matter that slows the running of some clocks (like the clock at the center of the sphere) but the fact that radiation from these clocks would be shifted to the red before that radiation could be received by any observer. This is the case because, in the system we have envisioned, there are two localities where the force due to gravity would be non-existent or insignificant -- the point at the center of the system for

159 See, P. Tipler, Physics, Worth Publishers, Inc. (1976), starting under section 16-6 at page 402.

160 See P. Tipler, Physics, Worth Publishers, Inc. (1976), within section 16-6 at Fig. 16-12 on page 404, for a graphic presentation of the variation of gravitation force with increasing distance from the very center of an object with a uniform density.

the reason noted above and the sphere of points located an "infinite" distance from that center."[161] Accordingly, if the Mossbauer effect is proof of General Relativity, then, extrapolating backward, a clock we have placed at the very center of a spherical shell would be the slowest clock in the system. Clocks at infinity would all "tick" at the same rate and this rate would be comparatively the fastest rate of any clocks located anywhere. The author has yet to see any discussion of this "paradox" in the General Relativity texts he has read. Moreover, even though both the clock at the center and the clocks at infinity are in areas where there is little or no gravitational "field", the pace of the clock at the very center of the system depends on the total matter in the system. If one doubles the mass of the spherical shell, then the clock in the field-free center of the system would move more slowly as measured by observers reviewing clocks at infinity. Thus, the clock at the center of the system has to know the entire mass of the system in order to know how fast to "tick" even though the mass of the system cannot create a net gravitational "field" at this central point.

A longer and more sophisticated criticism of the use of the Mossbauer effect to establish gravitational red shifts and thereby "prove" the validity of General Relativity can be found in L. Brillouin, Relativity Reexamined, Academic Press (1970), Chapter 6, especially at Sections 2 and 3, on pages 77-84. Mr. Brillouin's discussion is compromised to some extent by his continued reliance on negative gravitational potential energy but many of the aspects of his analysis ring true to the author.

Moreover, the fact of the gravitational "red shift" as confirmed by the Mossbauer effect leaves open the possibility of calibrating clocks to adjust for any time dilation factor. Consider, for example, that we might adopt a universal standard clock based on some particular atomic process such as a particular excited state to ground state transition of a

---

161 See again, P. Tipler, Physics, Worth Publishers, Inc. (1976), within
    section 16-6 at Fig. 16-12 on page 404.

shell electron of a standard chemical element.[162] General Relativity assumes that time must be measured to be synchronized with the frequency of this received radiation. If we know the frequency as emitted from the standard source, however, then we can apply a correction factor to the received signals so that, for example, if the received frequency is ½ of the emitted frequency, we can simply count two units of time for each crest or trough of the received radiation and have clocks that keep the same "time" at two radically different gravitational potentials[163].

---

162 See W.G.V. Rosser, An Introduction to the Theory of Relativity, Butterworth & Co. (Publishers) LTD. (1964) at section 1.2 beginning on page 1, and especially as page 2 continues to page 3

163 Note that without even a hint of awareness or uneasiness, astrophysics has already done precisely this and, moreover, presently attributes all red shifts to relative motion and not to differences in gravitational potential even though the author is not aware of any significant faction in the disciple that questions the established proof of gravitational red shifts. Thus, the author is continuously encountering discussions of astrophysics that assume a non-variable wavelength and frequency are associated with each emission line in the spectrum of each chemical element and many simple chemical compounds, further assume that neither the wavelength nor the frequency vary based on the element's or compound's environment or relative motion and still further assume that the clock rates of observers moving at different rates of speed or that are at different gravitational potentials can be adjusted to produce a coherent measurement of "time" for both observers. Those in the field, to be sure, do not explicitly rely on an adopted, universal standard to make their adjustments. The radiation sources are distant inanimate physical processes, not sentient observes applying any particular standard. Those in the field overcome this problem by simply (and quite naturally) assuming that the difference in wavelengths between different emission lines of a single element are no more affected by the element's environment and relative gravitational position than the frequency of any particular line is affected. Thus, astrophysics assumes that the pattern of a particular element's emission and absorption lines remains the same even if the pattern is shifted significantly to the red. On this basis, astrophysics assigns a velocity to a source of radiation by looking for a familiar series of absorption or emission lines (the Balmer series

In the same family as the Mossbauer Effect experiments are experiments and analyses that depend on nuances of the now existent "Global Positioning System." Since the GPS depends on orbiting satellites in relatively rapid motion (on the scale of objects that human beings typically deal with), electro-magnetic signals traveling material distances and upwards through gravitational "fields" and also involves distance measurements made based on time of travel for electro-magnetic signals with that time measured by clocks in a rapidly moving reference frame, many relativistic effects are evident in its operation. Most scientists seem to accept the proof of the effects inherent in the operation of the system as further proof of *General* Relativity. Most of these effects, however, are simply confirmation of the mechanics of *Special* Relativity coupled, to be sure, with the previously discussed

---

lines of Hydrogen, for example, which typically can be identified by their relative intensity and their frequency separation) and then calculating the red shift between the pattern of lines from this distant source and the similar pattern of lines in a sample at rest. Surprisingly, astrophysics does not even for a moment entertain the possibility that the red-shifted source may be stationary or be approaching but at sufficiently lower relative gravitational potential to produce the net red shift observed. Yet, if radiation from the central regions of one galaxy of roughly the mass of the Milky Way galaxy is received on the earth, which is toward the outer fringe of the Milky Way galaxy, that radiation would, in many cases, be red shifted more as it moves away from the source galaxy than it is blue shifted as it moves toward the Milky Way galaxy. Further, such a shift should be quite different depending on whether the radiation came from a source near the galactic center as distinguished from a source near the galactic edge. Indeed, red shifts of radiation that originates from within the Milky Way galaxy should be even more dramatic if the current belief that there is a "Black" "Hole" in the center of our galaxy is accepted as fact. If there is a "Black" "Hole" at the center of the Milky Way galaxy as the author believes is the current consensus, then radiation from near the center of our galaxy and headed toward the Earth would suffer an intense red shift during the course of its travel. The author has never heard any such shift discussed.

confirmation that there is a red shift in the travel of electro-magnetic signals away from a gravitating body.

As a fundamental proposition, however, a global *positioning* system is inconsistent with the most basic premise of relativity -- that there are no unique places in the universe. After all, the original purpose of the system was to assist the United States military in fixing precise locations and times on the earth so that commanders could coordinate the movement of their units on global battlefields and so that unique munitions could be delivered to unique locations at unique times to eliminate unique targets. Ultimately, the fact that one can make adjustments to the system to maintain its ability to locate unique places and times on the earth indefinitely and to bring unique ordinance into the presence of identified targets with precision means that the fundamental relativity principle of General Relativity cannot be true. Thus, although physics has abstracted the concept of an "observer" to one of a near "infinite" class of identical robots, this abstraction denies reality. The forward progress of physics bears witness to the significance of unique individuals providing unique insights over the march of a unique (and dare we say "absolute" although in only a single, master reference frame) scale of time. Every "observer" whose viewpoint has ever mattered, from Galileo to Newton to Einstein and all of the other great pioneers before, between and since, has been a unique being situated singularly within a span of time and confined to unique points at each and every definable time[164].

---

164 The author may seem, at first blush, to have descended into some nebulous philosophical aside but this is not the case. The uniqueness of every present and historical individual is an overlooked but nevertheless obvious "experimental" fact that proponents of relativity have never addressed. Of course, the discussion above makes a good case for the fact that modern physics and its proponents are too "sophisticated" to deign to consider the obvious. Current physicists are far too willing to accept complex solutions to simple problems and far too comfortable with logical inconsistencies. After these inconsistencies are buried under complex mathematical formulations written in cryptic and

c.   Comments on Perihelion Shifts as Proof of General Relativity

Moving beyond the experiments that rely on a gravitational red shift, we come to perhaps the oldest "proof" of General Relativity -- its apparent ability to explain a quirk in the motion of the planet Mercury. Like all of the orbits of all of the planets, the orbit of the planet Mercury is elliptical, at least in the first approximation, with the sun (or more properly the barycenter or center-of-mass of sun and planet Mercury system and probably still more properly the center-of-mass of the Solar System) at one of the foci of the ellipse. In each such orbit, there is a point of closest approach called the perihelion. In a classical, Newtonian analysis, this point of closest approach would be expected to be found at the same orbital position orbit after orbit. Historical observations of Mercury's perihelion, however, have demonstrated that the point of closest approach rotates in the plane of the orbit in the same direction that the planet moves. Classical gravitation could not explain this effect but General Relativity provided a formula for the amount of the rotation and the measured value and the formula value are reasonably close. Hence, General Relativity or something better than Newton's theory of gravitation must be used to explain Mercury's behavior. A strict classical analysis does not work. Arguably, then, General Relativity is thereby "proven".

Before commenting at length on the merits of this approach, the author would note the following concepts drawn from an article entitled "The Relativity Effect in Planetary Motions" by G.M. Clemence.[165] Mr. Clemence first observes that "Einstein's announcement of the general theory of relativity in its definitive form was immediately hailed by some astronomers as explaining a previously unaccountable discrep-

---

intensely abbreviated notations it becomes very difficult (perhaps even impossible) to challenge them despite their lack of validity.

[165] This article is taken from Tauber, Albert Einstein's Theory of General Relativity, Crown Publishers, Inc. (New York, 1979), within Part III beginning on page 126.

ancy between the observed and theoretical motions of [Mercury]". He continues that "[o]thers were, however, intuitively opposed to relativity, and they directed attention to a small discrepancy yet remaining as evidence that the theory of relativity could not be correct: the relativists contended that the small remaining discrepancy was due to errors either in the observations or in the classical theory of the motion"

Mr. Clemence next observes that the measurement of the orbit of Mercury and thus of any discrepancies in that orbit is a very complicated affair. According to Mr. Clemence observations of the position of Mercury are difficult because they must be "made in the daytime, near noon and under unfavorable conditions of the atmosphere". As a result, these observations are subject to "large systematic and accidental errors" and these errors are increased due to the shape of the visible disk of the planet. Mr. Clemence continues by noting that Mercury's "path in the Newtonian space is not an ellipse but an exceedingly complicated space-curve due to the disturbing effects of all of the other planets" and "[t]he calculation of this curve is a difficult and laborious task". He observes that different results have been obtained by different computer calculations. He further comments that observations of Mercury cannot be made in a Newtonian frame of reference, but, instead, "are referred to the moving equinox, that is, they are affected by the precession of the equinoxes." According to Mr. Clemence, "the determination of the precessional motion is one of the most difficult problems of position astronomy, if not the most difficult". Thus, it appears that the identification of Mercury's orbit and the identification of anomalies in that orbit of Mercury are daunting tasks and that reasonable men may differ as to the extent to which the orbit and its anomalies have been quanitified.

Further in Mr. Clemence's paper, he concedes that "the observational material is so extensive and the methods of analysis so complex that it is not practicable [in his article] to present any evidence that [would] enable the reader to form an independent judgment of the errors involved". As far as specifics concerning the observations required as part of the process, Mr. Clemence indicates that:

> The observations of Mercury are of two different kinds: observations of its spherical coordinates on the celestial sphere when it is on the meridian, and observations of the time at which its disk is tangent to the disk of the sun when Mercury crosses the face of the sun. The meridian observations extend from 1765 to 1937 and number about 10,000 in each coordinate. Observations of 17 transits have been used, extending from 1799 to 1940.
>
> The observed coordinates are not discussed directly, but instead the small differences between the observed coordinates and those calculated from a theory of the motions are used. Each of these differences gives rise to an equation of condition, the unknown quantities being corrections to the constants used in the calculated coordinates. These equations are collected into groups extending over about ten years each and solved by the method of least squares. The number of unknown quantities is twelve, one of them being the correction to the assumed, or tabular, position of the perihelion. In principle, a number of corrections at successive epochs to this assumed position of the perihelion are obtained, and the sum of the corrections gives the correction to the assumed motion of the perihelion. The procedure followed with the transits of Mercury is much the same, except that the whole series of transits furnishes only two equations of condition because transits can occur only in two narrow regions of Mercury's orbit. These two additional conditions are imposed on the final results of the meridian observations, and another adjustment is made by least squares.

Despite what the author perceives as the dire forebodings in much of the text above and the complexity of the observations that are the basis for the analysis, the final conclusion of Mr. Clemence's article is that the theoretical relativity effect on Mercury's perihelion is 43".03 ±.03 and the value obtained by subtracting all other known effects from the total observed motion is 42".56±.94 so that these two figures are in remark-able agreement. From this, it is perhaps understandable that many have confidence in General Relativity, and, of course, there are few with the skills to mount a complete critique of the work that has been done and likely none who would, at this point, prepare this critique to support the heresy that there are problems with General Relativity. The author cer-tainly must confess doubt as to his skills in any such endeavor and, in any case, does not have the time to attempt to develop and apply such skills.

Nevertheless, the following points seem germane to the issue. First, there are a large number of assumptions made in assessing the theoretical motion of the planet especially perturbations caused to Mercury's ellipse by the gravitational influence of the other planets in their orbits and to the shape and mass distribution of the sun itself. Those testing the effectiveness of General Relativity use Newtonian concepts to establish the mass of these objects, their shape and their influence and such a process seems to assume an accuracy of the Newtonian approach that can no longer be justified. They do not use General Relativity to improve their theoretical calculations, of course, because the mathematics of that theory are so intractable.

To highlight the significance of such assumptions on the quality of the proof, the author would note that the same work that includes the article quoted at length above also includes the following text from a follow up article[166]:

> What is the status of Einstein's three tests? The effect on the orbit of Mercury remains a most impressive test of General Relativity. Its impact in 1915 was dramatic because the pre-dicted motion of 42.9 arc-sec/century so closely matched the observed anomaly, which had been corrected by then from

---

166 See, G. Tauber, Albert Einstein's Theory of General Relativity, Crown Publishers, Inc. (New York, 1979), within Part III beginning on page 143 "The Status of Einstein's Three Tests and of Shapro's Time-Delay Test.

Leverrier's original estimate of 38 arc-sec/century to 43.3 ± 0.3 arc-sec/century. A weak point concerns the influence of the sun on the orbit of Mercury. If the sun is oblate -- that is, flattened at the poles like the Earth -- there will be a Newtonian perturbation of the perihelion motion from it as well as from the planets. Originally any such effect was thought to be less than 0.05 arc-sec/century, but in 1964 R.H. Dicke, following some earlier work with C. Brans on the Jordan scalar-tensor theory of gravitation, suggested that the inside of the sun might be rotating up to ten times faster than its surface causing an oblateness large enough to account for 10 % of the Leverrier anomaly and a consequent 10% discrepancy with Einstein's theory. Careful measurements by R.H. Dicke and H.M. Goldenberg on the optical shape of the sun in 1968 seemed to confirm this. More recent observations by H.A. Hill and R.T. Stebins disagree and appear to suggest that the optical shape of the sun fluctuates with time and cannot be taken as a reliable measure of its mass shape. Without attempting to take sides in this controversy, one may justly remark that the mass distribution of the sun has never been measured directly, and that the hypothesis that the sun has a rapidly rotating inner core is an eminently plausible one.

Further, according to Brillouin,

The advance of the perihelion of Mercury (43 seconds per century) was hailed as a wonderful check with a theoretical prediction of 42" 6, but here again let us refer to Chazy (1930) who found a number of other examples in the solar system where Einstein's predictions conflict with experiments. It is hard to believe seriously in a coincidence of less than one second for Mercury, while so many other examples give large errors and even opposite signs! Let us here candidly admit that there must be many other unknown factors involved. The computations of Chazy refer to the motions of perihelions of four planets and similar motions for a number of satellites orbiting around planets (e.g., the moon). Errors of at least five seconds per century seem to be the inevitable limit in these very difficult computations. Einstein's theory yields about 1/6 of the advance of the perihelion of Mars and practically nothing for Venus. Let us add that Dicke's discovery of the oblate shape of the sun leads to perturbations that definitely destroy the agreement about Mercury. The question cannot be considered completely settled.

See L Brillouin, Relativity Reexamined, Academic Press (1970), Chapter 7, where the quotation is taken from page 99.

The author would hasten to add that the approach to gravity that he suggests would have an influence that somewhat mirrors solar oblateness. Applying the author's approach and ignoring the influence of "objects" outside the Solar System and thus focusing only on the sun and its planets, we would need to impute potential energy into the space between the sun and the planet Mercury based on the positions of the other planets at the time we are attempting to calculate its orbit. The quantities of potential energy that we would expect to find within the orbit of the planet Mercury would have the following mass equivalences:

Planet	Mass	Negative Potential Energy at the Sun's Surface (with figures converted to Kilograms)[167]	Negative Potential Energy at Mercury's Radius	Negative Potential Energy Imputed Within Mercury's Radius (with figures converted to Kilograms)
Mercury	3.18×10^{23}	6.75×10^{17}	8.11×10^{15}	6.67×10^{17}
Venus	4.88×10^{24}	1.03×10^{19}	1.24×10^{17}	1.02×10^{19}
Earth	5.98×10^{24}	1.27×10^{19}	1.52×10^{17}	1.25×10^{19}
Mars	6.42×10^{23}	1.36×10^{18}	1.64×10^{16}	1.35×10^{18}
Jupiter	1.90×10^{27}	4.03×10^{21}	4.85×10^{19}	3.99×10^{21}
Saturn	5.68×10^{26}	1.21×10^{21}	1.45×10^{19}	1.20×10^{21}
Uranus	8.68×10^{25}	1.84×10^{20}	2.22×10^{18}	1.82×10^{20}
Neptune	1.09×10^{26}	2.32×10^{20}	2.78×10^{18}	2.29×10^{20}
Pluto	1.40×10^{22}	2.97×10^{16}	3.57×10^{15}	2.94×10^{16}

From this table, it is clear that the potential energy within the orbit of Mercury that would be attributable to the most massive of the plants -- Jupiter, Saturn and, to a much lesser extent, Neptune and Uranus -- is modest but not trivial and is far more significant than the potential energy attributable to the remaining planets. Whether the movement of the perihelion of the planet Mercury could be reconciled to an approach that

---

167 Using the classical potential energy formula,

$$(\text{Potential Energy})_{\text{Gravitational}} = - \frac{G(M_1)(M_2)}{(r_{1,2})}$$

From J. Jewett, Jr. and R. Serway, Physics for Scientists and Engineers, 6th Ed. (Thompson Brooks/Cole 2004), Section 13.6 with emphasis on the language at page 404 with the orbit of Mercury taken at its average distance of 2.43×10^{6}, the Gravitational constant at 6.673×10^{-11}, and the result divided by $8.987\ 551\ 79 \times 10^{16\ m\ 2}/sec^2$ so that the energy figure is converted to a mass equivalence.

places potential energy within its orbit as implied by the locations of the other planets will take an analysis that the author has neither the skills nor the time to undertake. Perhaps now that there is something to look for, however, someone with the time and skills will undertake the effort. The author takes comfort, however, in the fact that the data so far has been "processed" to explain only the advance of Mercury's perihelion. If the author is right, then the gravitational effects that would be experienced at the position of the planet Mercury due to the gravitational influence of the potential energy implied considering the locations of Jupiter and Saturn would be radically different than the simultaneous gravitational influences of the implied potential energy of Jupiter and Saturn at the locations of all other celestial bodies. Accordingly, the data can be made to explain the behavior of Mercury only at the expense of an inability to explain the behavior of any body moving other than in synchronization with Mercury. Indeed, objects that have a periodicity that is out of synchronicity with Mercury would demonstrate effects that have the opposite sign as those of Mercury. The comments of Mr. Brillouin quoted above, of course, indicate that there are some bodies whose perihelions suffer effects that are quantitatively different in sign than those of Mercury. If the author is right, that would be expected.

11.   Widely Accepted Concepts and Constructs that Are Erroneous

Given the discussion above, there are phenomena that have been suggested as existing based on the precepts of General Relativity that the author believes will never be observed and constructs that are popularly employed in the application of General Relativity that the author submits may no longer be used. As to the former, there is one very popular phenomena that is now widely accepted as an established fact[168] but that cannot exist if the ideas in this work are correct. This accepted but non-existent phenomena is that of a "Black" "Hole." Similarly, as suggested above in the listing of erroneous holdover concepts from Newtonian gravitation, the constructs of a gravitational "field" and of a "test" mass that can be used in the construction of such a "field" must both now be abandoned in the development of a new rigorous theory of gravitational attraction. The comments which follow further explain why.

---

168 Although there now appears to be wide spread agreement as to the existence of what have popularly been described as "Black" "Holes", the author feels constrained to point out that such phenomena cannot, by definition, be "observed". Light and other forms of electromagnetic radiation cannot escape from these phenomena, so that, as a result, they cannot be detected by direct observation. Instead, all that can be done is to infer the existence of such phenomena based on the behavior of things that can be seen. If the theory of gravitation that supported the inferences in favor of these phenomena falls, so too must the inferences from that theory.

a. The Failure of the Concept of a "Black" "Hole"

The prior discussion concerning gravitational force associated with the potential energy stored in a gravitational "field" suggests problems in associating the now popular concept of a "Black" "Hole" with any real galactic or sub-galactic distribution of matter. Such problems probably should not be a surprise[169]. The fact that 90 percent or more of the

---

169 As far as the author can determine, the theory that describes the birth of the universe as a "Big Bang" remains dominant in cosmology. A material feature of this theory is a brief period of "inflation" after the "bang". Accordingly, the established approach to cosmology suggests that all of the inescapable black holes that are subsets of the current universe -- together with all of the stars, planets, dust and radiation in the present universe as well -- were nevertheless able to escape from each other in this initial bang. Moreover, all of these intensely massive things (or their predecessors) apparently were able to move faster than the speed of light during the period of "inflation". The author has never seen convincing reconciliations of these ideas and modestly submits that these ideas are glaringly inconsistent. Further, the author is continually surprised to read explanations attributing intense energetic phenomena to the affects of black holes. Conceptually, a black hole is an omni-directional vacuum cleaner. To peer searchingly into the murky depths of such an apparatus, discern the equivalent of a cannon ball rapidly rising *from those depths* and then to declare the movement of this cannon ball as proof that the apparatus is, indeed, the vacuum cleaner as described requires intellectual leaps far beyond the author's capacity. Note that an obvious alternative explanation to the concept of inflation is that the measurement of time here on Earth is slowed due to its motion relative to the center of mass of the universe. Time dilation is real. Such a realization would make it unnecessary for there to be a discontinuity in the rules of physics as the concept of inflation implies. Indeed, using the author's approach, one could develop a "steady state" but evolving and apparently expanding universe to fit many ideas that are part of the "Big Bang" theory. One could model the universe as a giant fluid ball with the "fluid" consisting of a master evolving electromagnetic field with a stationary center. Charged particles represent voids or bubbles in

mass associated with typical hypermassive galactic systems is unanticipated and unexplained under the existing understanding of gravitation should be a warning that the confident use of that existing understanding in modeling smaller, starlike, but still hypermassive objects (i.e., stars that purportedly have relativistic escape velocities) has likely produced anomalous, unphysical results as well.[170]

The author's analysis of the theoretical basis for such objects confirms that this warning should have been taken to heart -- and that the problem ultimately lies with the erroneous convention as to the zero-point of gravitational potential energy outlined in one of the opening sections of this work. Thus, in the current theoretical analysis leading to the supposition that there are such objects as "black" "holes," the mass of each discrete "object" is enhanced based on a perceived need to incorporate gravitational potential energy into the object's construction. More formally, the assignment of mass to an "object", starts with the concept of a perfect fluid and assumes that this perfect fluid is prevented from collapse by internal pressure resisting further compaction by gravity.[171] The "mass" of this fluid assembly, then, is assumed to be

---

that fluid. Just as bubbles in a fluid rise to the surface, all objects that consist of charged particles might very well move away from the dense center and toward the periphery. Complex mutual induction features of electromagnetic fields might very well cause the bubbles to clump as they move away from the universal center. Inflation, then, would represent the impact of the rising motion on the local clocks we use for observation. Of course, musings on what could be the nature of the universe such as these comments hardly represent exact science if in fact such musings may be called science at all.

170 The descriptions the author has read of the lack of parallelism between the experience of an observer that has passed the Schwarzschild radius and the appearance of that experience to an outside observer leaves the author all the more suspicious that the theoretical apparatus which predicts these differing experiences deviates from reality.

171 See T. Frankel, Gravitational Curvature (An Introduction to Einstein's Theory), W. H. Freeman and Company (1979) at 26-31.

the volume integral of its rest density coupled with the volume integral of its potential energy of compression.

To determine the potential energy of compression, one assumes that it is equal to (but opposite in sign from) the volume integral of gravitational potential energy. The gravitational potential energy that is selected by those in this discipline, however, is defined as the work done in moving a unit mass from its current location out to an infinite point rather than the work done to move a point out from the center-of-mass of the system to its current location.[172] Thus, the gravitational potential energy currently ascribed to an object is the energy required to cause the assembly to do something unnatural -- to explode into an infinitely distant and dilute spherical shell. The final, low-energy state of a stable gravitating system, however, is not this infinitely distant dilute shell but an infinitely dense compact point. The "gravitational" potential energy that those in this field attempt to identify as a source of gravitational force is, in reality, then, the kinetic energy that the components of a gravitating object would have carried to their current locations in the course of a hypothetical fall from infinity rather than the additional kinetic energy (and thus current potential energy) that could be obtained if the components were allowed to fall farther and all the way into the low energy state of a compact point mass. In other words, in its effort to incorporate potential energy into its analysis, General Relativity takes the positive portion of classical gravitation's fixed sum of kinetic plus potential energy and re-labels it as "gravitational" potential energy. The real gravitational potential energy is the other, traditionally negative term.

As a result of this conceptual error, the potential energy that *actually* could be realized in a gravitational collapse has nothing to do

---

172 See, T. Frankel, <u>Gravitational Curvature (An Introduction to Einstein's Theory)</u>, W. H. Freeman and Company (1979) at 29 to 31 and also 22 to 25. Note, in particular the language at page 22: "[t]he 'gravitational potential' at x should be the work done in moving a unit mass from x to infinity (note that physicists usually call the negative of this quantity the potential)."

with the "gravitational potential energy" presently assigned to a given gravitating structure by General Relativity. The gravitational potential energy that *really* is "located" *inside* an object is the energy that would be realized if the object is allowed to collapse into the tightest permissible space. As detailed in the paragraph above, however, the amount that is actually incorporated under the current analysis represents the potential energy that would now be stored in our object had its components started at infinity, been brought to a stop at their present location by the process that formed the object and have lost none of that potential energy in the process of the object's formation.

The nature of the error explains a common assertion with respect to all objects: that every object has a level of compaction -- a final radius -- that would cause each to be a "black" "hole." Such a conclusion is precisely what would be expected if the wrong potential energy term is incorporated into the object's construction. A diffuse mass distribution, even of trivial aggregate initial rest mass, inevitably produces an object of infinite mass if one allows the potential energy that is accumulated as the distribution collapses toward a dimensionless point to augment the initial rest total and if one does not include the mass-equivalence of this energy in initial calculations. As each shell that makes up an object collapses, it accumulates the potential energy that would have been required to move that shell out beyond the space through which it has just collapsed. This potential energy did not previously exert a gravitational pull because it was "negative". Now that it has reappeared as positive mass, it and the original positive mass are able to accumulate still more positive mass by falling further, converting still more "negative" potential energy into more positive mass and so forth leading to an explosive growth of "mass". If one starts with a system such that a given quantity of fuel would only permit a defined object to reach a specific distance and then posit the same object at a still greater distance, then one has an infinitely massive system and its collapse yields an unphysical object -- a "black" hole".

Indeed, the current supposition that every traditional object can become a black hole if its radius becomes small enough ultimately is a simple corollary to the pre-relativity precept that a single attractive force like gravity cannot produce a stable system[173]. As the distance between two gravitating sub-parts of a system tends to zero, the force between those two sub-parts tends to infinity and the work required to separate them from each other (which is the infinite force times a finite distance) would also tend to infinity. Since General Relativity is constantly transmuting this ever growing potential energy into traditional gravitating matter in the course of a collapse, the collapse of any system inevitably results in an infinitely massive object.

Interestingly, then, a problem that classical physicists all understood and knew must be avoided -- indeed, the problem that almost surely led these physicists to select the zero-point of gravitational potential energy at infinite separation as discussed previously -- has become a feature confidently incorporated into the dominant theory of gravitation. The consequences of ignoring this problem have simply been papered over through application of the intractable mathematics of General Relativity. The author frankly has come to wonder if there is anyone competent to use the elaborate mathematics that he repeatedly has had to wade through in an effort to understand what those in the field of gravitational physics are really doing. If this work has value -- and it likely does if you have read through it to this point -- it is clear that the overwhelming majority of even the most highly educated physicists do not have a real understanding of the basic concepts buried under the mathematics they employ[174].

173 See E. Purcell, Electricity and Magnetism, (McGraw-Hill Book Company 1965)¶1.6, especially at the bottom of page 15 where this point is made in the discussion of an electrical system -- a salt crystal. See also, M. Born, Einstein's Theory of Relativity, Dover Publications, Inc. (1965) at 221, just below the text at that page referenced in a prior footnote.

174 There is a suggestion that, when Albert Einstein offered his paper on General Relativity to a publisher, he "warned them that there were not more than twelve persons in the whole world who would understand it." See S. Goldberg, Understanding Relativity, Birkhauser Boston, Inc.

(1984) at page 312. If the author is correct, the number likely is less than twelve, and, surprisingly, Albert Einstein was not one of them. It is clear to the author, then, that the select few who really can use the complicated mathematics that adorn so many current advanced books on physics should reconsider their mission -- it is inappropriate to write for the hypothetical select twelve as many currently do. Rather, those that really can understand such complex mathematics need to spend their time translating their understanding into simpler language for the rest of us. The author has cited, again and again, the works of a few authors -- W.G.V. Rosser and E. Purcell, for example -- because they have largely succeeded in making complex ideas accessible both mathematically and intuitively. In contrast, the many scientists who have attempted to explain Relativity with complex mathematics using abbreviated presentations and over-complicated, over-generalized abstractions -- the "Einstein Summation Convention" for the presentation of complex relations, "generalized" or "curvilinear" coordinates in lieu of Cartesian ones for presentation of many problems, to give but two examples -- have saved (in the first case) or wasted (in the second) a great deal of paper at the expense of imparting any real understanding. If the author is correct, they have produced nothing of value throughout their entire careers. Indeed, to the author, it appears that many who make their living in the sciences are -- perhaps only subconsciously -- more interested in those careers and in protecting their livelihoods (and those of their friends in the mathematics department) as teachers and their privileged status as an elite rather than in really advancing human understanding. Many of their books are texts complete with exercises for students and many read like the bridge column in the local newspaper -- given enough time and a determined analysis, the constraints they and the bridge columnist have identified but not specifically stated can be discovered, the terminologies they both have adopted can be understood and the conclusions they both have reached can be seen (for the most part) to be the correct ones. To be sure, this is no complaint about the quality of the work of the bridge columnist. The columnist creates an amusement and it is an amusing part of the "game" of solving the problem to discover what the columnist has left out. The search for truth regarding the physical universe, however, is not a game. Satisfaction comes with knowing, much more than solving, at least, when it is known that the problem has already been solved. Accordingly, it is time to stop leaving the "proof" to the reader, to stop leaving the development of basic concepts to "exercises" the reader is to perform, to stop the proliferation of needless mathematical prerequisites, abstractions and terminologies and to include each and every step in a proof of an assertion from the first step to the last before that assertion can be deemed to be proven. Moreover, when the principle relied upon in the proof is drawn from mathematics, the proof should include proof of the mathematical elements in even more rather than less detail than the physics. As it currently stands, in order to be able to understand most modern texts on General Relativity and many on Special Relativity, one would have to dedicate years to the study of abstract mathematics often generalized to infinite dimensioned vector spaces using complex "tensor" presentations -- typically introduced using the heavily abbreviated "Einstein Summation Convention" -- in order to have the simple ability to understand what those in physics think they know about the three spacial and one time dimensioned universal space in which real people live. To be brutally frank, the author sees the story of General Relativity as proof that *all* seeking to understand its features were exhausted in their effort well before they developed a proper conception and, worse still, many became afraid to question those who were not exhausted but rather misled. Although pride in accomplishment of difficult tasks is a natural emotion, we must fight the temptation to criticize those who are honest enough to admit a lack of understanding and need a little help. Often, it is when the select few with a more complete grasp of the situation seek to provide such help that they themselves obtain new insights and understanding. Further, we must resist the temptation to believe too much in any one person's powers of reason, be it Albert Einstein or our own. History is littered with now discarded concepts -- the Philosopher's Stone, the fountain of youth, lost cities of gold, witches, therapeutic bloodletting, etc. -- that thrived because individuals did not dare to question what was false for fear of drawing the sneers of supposed experts. A real expert, of course, is not someone whose credentials entitle their positions to acceptance but, rather, someone who understands and can explain his or her position so well that a reasonable person cannot help but be persuaded. It should be remember, too, that most of those who have applied their lives to advancing the current state of General Relativity either work for government-sponsored educational institutions or accept substantial funds from the government to assist in their researches and the writing of their texts. If we as a people are paying for their work, their work is not complete until they can satisfy us that the money was well spent. They cannot simply present their credentials, couch cryptic concepts in hieroglyphics and insist that whatever they have done is brilliant and that we just have to take their word for it. Indeed, to illustrate

Further on the author's concerns as to the validity of the concept of a "black" "hole", the author believes that the presence of a "black" "hole" within a galactic distribution of matter implies that gravity is not a conservative force and thus would not be a radial force. This is the case because, in a central force "field", the energy associated with a particular location should be independent of the path by which the particle reached this location. This notwithstanding, it is possible to move objects from one side of a "black" "hole" to another by means of paths that do not go within the "Schwarzschild radius" of the object yet it would not be possible to start at the same point and end at the same point by passing through the "Schwarzschild radius" even if one does not actually collide with the singularity that is the hole. Thus, it is possible to find a paths that avoid singularities but that are still not equivalent.

The author would further note that the existence of "black" "holes" within the framework of General Relativity highlights a significant disparity between the treatment of time, distance and velocity in General Relativity as opposed to Special Relativity. In Special Relativity, there is parallel treatment between the perception of differences in time and the perception of differences of distance in the direction of relative motion. Indeed, the perception of these differences is, at times, described

---

the depth and breadth of the problem, it is well to note that the essence of what is really know about electromagnetism, when separated from materials science, can be displayed on a tee-shirt and taught in less than a week. This notwithstanding, the abstract mathematics one currently needs to be able to understand what physicists are now writing about electromagnetism would take a decade or more of intense study. In the final analysis, the author believes that all of this mathematics is a cloud that obscures rather than enlightens. Indeed, the story of the Tower of Babel comes to mind -- the mathematics in use has become so complex that people cannot work together to build a realistic understanding of the universe. He who would challenge this assertion needs to be able to explain why the thousands who have worked in gravitational physics since 1905 have never recognized the falsity of any analysis that uses negative energy figures and why the individual who finally recognized this glaring error is not a physicist at all.

as a rotation of the orthogonal axes of the "time component" and of the "distance along which there is motion in one frame of reference component" through an angle with a tangent equal to iv/c.[175] The end result is that the "time" and "distance along which there is motion" axes each have precisely interlinked components when viewed from another uniformly moving frame. Within this framework, then, there is a maximum difference in the rate of change of distance with time and also a maximum rate of change of time with distance. There is no such parallelism in General Relativity. In General Relativity, time goes more slowly the deeper within the gravitational "field" the clock is located and the rate differentials remain the same with the passage of time in both frames. The time difference between a point deep in a gravitational "field" and a point at the periphery of a gravitational "field" will therefore grow linearly and this difference will tend toward infinity so long as we contemplate the possibility of infinite time. The spacial distance between these points, however, may remain static or even decrease. In General Relativity, then, there is an inherent lack of parallelism between differences in the rate at which spatial coordinates can change and the rate at which time coordinates can change[176].

---

175 See W.G.V. Rosser, An Introduction to the Theory of Relativity, Butterworth & Co. (Publishers) LTD. (1964) at section 6.3 beginning on page 262.

176 To the author anyway, the process of combining three spatial coordinates and a single time coordinate with the concept of an invariant "interval" works only so long as a light ray must "chase" a moving observer along the path of increasing separation in order to make its presence known. While this certainly makes sense for constant motion in a three dimensional space, and for acceleration in three dimensions along the path of separation, it makes no sense for a path that returns upon itself such as a circular orbit. A constant series of spherical pulses issued from the center-of-mass of a system directed to an object rotating around that center-of-mass in a circular stable orbit will all travel the same distance and will all have the same time of travel. Such a traveler's clock should run slow but the observer would be able to adjust the clock settings continuously based on an agreed convention with a center-of-mass clock (which, as discussed above, keeps a "true" time). The convention

Also, the formulation of the radial component of the line element used in the analysis that leads to the assumption that there are black holes starts with the standard 3 spacial dimensions and 1 time dimension. To meld these 3 distinct dimensions into a 4-dimensional space-time, the time dimension is typically represented by a complex number, *ict* so that the line element has 4 equivalent components X_1, X_2, X_3, and X_4.[177] But the components are not really equivalent. Thus, for example, if we assume that the X_1, X_2 and X_3 components are each ½ way to infinity, the X_4 component would have to have gone to "infinity and beyond" to borrow a cartoon-ish phrase from a popular children's movie. The author would not be surprised to find that the point at which the 4th dimension reaches infinity (and the spacial dimensions are, on average 1/3 of the way there) is, in fact, the Schwartzchild radius[178].

Finally, the discussion of "pressure" in the derivation of "black" "holes" highlights a radically different concept of "pressure" in areas

---

would be to ignore any "gravitational" red shift in the incoming signals and simply count the number of messages sent or rely on information content of a message (i.e. a message that indicates that it is now "10:00 followed by a message 1 hour later in the center-of-mass frame that says it is now 11:00"). Indeed, we could always synchronize our updating signal to the rotating object so that the updating signal would always follow precisely the same path and intercept the orbit of the rotating object at precisely the same angular position each time it is sent.

177 See again, W.G.V. Rosser, An Introduction to the Theory of Relativity, Butterworth & Co. (Publishers) LTD. (1964) at section 6.3 beginning on page 262.

178 The author would add that, if we assume that each of what are traditionally viewed as "spacial" components (X_1, X_2 and X_3) are 3/4th the way to infinity, the X_4 component would have to have gone to "infinity and beyond" twice and we would expect there to be a second problem point in the line element used in General Relativity's analysis -- the point at which the spacial dimensions are, on average 2/3 of the way to infinity. According to Brillouin, there is such a second problem point. See L Brillouin, Relativity Reexamined, Academic Press (1970), Chapter 4, Section 7 at pages 52-53.

close to the center-of-mass of a hypermassive system. Pressure as presently understood is the result of the average number and kinetic energy of molecular impacts. In an area close to the center-of-mass of a hypermassive system, the difference in gravitational potential over a small distance would be quite substantial and objects of trivial rest mass such as molecules could develop immense kinetic energy in falling through a very small distance. One would therefore expect the center of such a system to contain very little traditional matter. Moreover, the random motion that we normally associate with the "heat" of this traditional matter would be able to do substantial work. Our conception of gravity, developed toward the edge of a galactic system, therefore, may provide us with very little insight into the behavior of gravitating objects at the center of a galaxy or systems of galaxies.

b. The Need to Abandon the Field and Test Mass Constructs in a Proper Theory of Gravitation

As noted previously, the gravitational force associated with the potential energy stored in a gravitational "field" has lead the author to abandon the ancient concepts of a gravitational "field" and of a "test" mass that can be used to construct such a field. This is the case because the "field" concept and the use of a "test" mass to identify it cannot be reconciled with conservation of mass/energy.

Thus, consider a system that includes a central object with an escape velocity of .999999999 percent of the speed of light and four pairs of additional and identical test objects (our candidates for potential real "test" masses) each of mass "m". Each test object is assumed to have a trivial mass as compared to the central object. A given pair of test objects will be split and placed at the same distance from the central object but on either side of it (so that the center-of-mass of the system never moves and provides a constant reference frame for the computation of the mass of the overall system). These pairs are inserted at four distances from the central object: (a) the distance which will result in an impact velocity of .95 percent of the speed of light; (b) the distance which will result in an impact velocity of .9995 percent of the speed of light; (c) the distance which will result in an impact velocity of .9999995 percent of the speed of light; and (d) the distance which will result in an impact velocity of .999999995 percent of the speed of light.

If one allows these "identical" pairs of "test" objects to fall into the central object, each pair will arrive at the central object with a different mass. The objects that started at the distance generating a final velocity of .95 of the speed of light will each have a mass of $3.203m$ when they complete their falls. The objects that started at the distance at which their speeds reached .9995 of the speed of light will each have a mass of $31.627m$ when they have fallen. The objects that started at the distance at which their speeds reached .9999995 of light speed will each have a mass of $1,000.000m$ at impact. Finally, the objects that started at the distance at which their speeds reached .999999995 of the speed of light will each have a mass of $10,000.000m$ when their journeys are done.

From this analysis, it follows that not one of the test objects can properly be considered to be a "test" "mass" since the amount of mass each adds to the system depends on where the test object begins its testing.

The fact that there are no "test" masses compels the conclusion that the "field" conception of gravitation also must be abandoned. To illustrate why, let us assume that the objects inserted at the relevant distances outlined above are identical to the central object rather than mere test objects. Consistent with the current concept of a gravitational "field", they will experience the same series of accelerations and reach the same final velocities as the test objects do. In the most extreme of the examples, a system that begins with three times the mass of the central object and with a pair of the threesome inserted at the distances suggested in example (d), above, ultimately comes to have a mass 20,001.000 times that of the central object upon collapse.

It is easy, of course, to take 5 such objects with 4 at the target distance and generate, using current gravitational concepts, a collapsed object that has a mass of 40,000 times the mass of a single object with an escape velocity of .999999995 percent of the speed of light. One can add any number more of these objects at the appropriate distance and, for each, add another 10,000 such masses to the system upon collapse. Note as well that, in this example, the author has addressed only the interactions between the central object and each distant object and has ignored the interactions among the distant objects. Even this example, then, substantially understates the rapidity with which the current approach to gravitation requires a deviation from the fundamental mass/energy conservation principle.

Based on the above, there cannot be a gravitational "field" and one cannot use test masses as proxy to determine the behavior of other, larger masses when inserted at a point in space.

12. Suggestions for Development of a Coherent New Theory of Gravitation

Throughout this work, the author's view point and statements have largely been negative. The author initially asserted (and has hopefully proven) that there are flaws in a variety of very basic current assumptions regarding gravitating systems. These erroneous assumptions are a legacy of the Newtonian understanding of gravitational phenomena and they are flawed because they are inconsistent with the mechanics of Special Relativity. These have gone undetected because Albert Einstein moved too quickly to develop an approach to gravitation after his 1905 paper introducing the concepts that are the core of Special Relativity. He should have -- but did not -- systematically review the established concepts of mechanics and gravitation to check those concepts against the new mechanics he helped to develop. Had he systematically reviewed all existing concepts of mechanics and gravitation, he would have realized that the assignment of zero to the gravitational potential energy of points at an infinite distance and all the concepts that go with that assignment could no longer be maintained.

The author has undertaken some portion of this effort, has identified quite a number of erroneous concepts grounded in pre-20th century thought that remain part of the gravitational canon today and has further endeavored to trace the impact of these flaws and related conceptual inconsistencies through the great mass of the mythology of General Relativity. Given the discussion above, the author believes that it will be difficult for General Relativity to survive at all and it certainly will require substantial revision in any event. There simply is no way that the existing experts in the field will be able to defend their current, object-based approach or the use of "negative" energy quantities in any rigorous gravitational analysis or the historical convention that systems that are full of gravitating energy are nevertheless empty of any extra gravitational influence because these experts' peers -- in reality, their ancestors -- agreed that this was so. Absent further effort, we are left without a definitive approach to the oldest and most basic of all of the "forces" of nature. The author, then, would like to conclude this work with his blueprint for the construction of a proper understanding of gravitation and a more systematic understanding of the concept of energy.

To begin the exercise, the author would first state his view that there simply cannot be an independent, attractive force existing between all "mass" and all "energy" in the universe if the concepts and mechanics of Special Relativity are respected. Were there to be such a force, the mass equivalence of the energy required to overcome the force between two objects would operate on the objects and on itself the moment one does work to separate the objects leading to insoluble infinities. As classical physicists well understood, a single, attractive force is unstable and would inevitably lead to collapse[179]. There cannot, then, be a "force" of gravity as either Newton or Einstein understood it. Albert Einstein, of course, worked for years on this problem without success -- his "Special Relativity" paper was published in 1905 and his paper on General Relativity was not published until 1916 -- and never really addressed the problem directly. He simply believed he could finesse the problem by developing a theory of gravitation based on the principle of equivalence. As we have discussed at length above, this effort failed because the principle of equivalence on which it is based is only an approximation appropriate to gravitating systems of modest aggregate mass and limited separation distances.

---

179 See E. Purcell, Electricity and Magnetism, (McGraw-Hill Book Company 1965)¶1.6, especially at the bottom of page 15 where this point is made in the discussion of an electrical system -- a salt crystal.

The absence of a "force" of gravity notwithstanding, there clearly is a property of the universe that tends to draw together any two "objects" in that universe. Further, Newton's approach to gravitation clearly describes the application of this property within the Solar System to a very close approximation. Clearly also, this property operates locally but is the result of distant objects sited in their "current" position as measured -- using the Lorentz transformation equations -- by the observer experiencing the consequences of the existence of this property. Nevertheless, this property must not depend *directly* on the current locations of the two separate objects as measured simultaneously in the frame of reference of the observer. Were transmission of the attraction to occur instantly, there would be problems with causality as made clear in Einstein's writings.

To construct a theory within these constraints, the author suggests that there must be reliance exclusively on analyses grounded in electro-statics and electro-dynamics. A scientist like Einstein, operating 100 years ago or so, might have reasonably attempted to build a theory of mechanics using "rigid" rods and ideal "clocks" but we now know that such idealizations are not defensible. There are no "rigid" objects in the universe[180] and the reliance on electrically "neutral" objects and the periodic behavior of some "mechanical" systems such as pendulums or even the vibrations of crystals fits 17th, 18th and even 19th century concepts but not current ones. There are no stable "neutral" objects[181]

---

180 W.G.V. Rosser, <u>An Introduction to the Theory of Relativity</u>, Butterworth & Co. (Publishers) LTD. (1964) at section 5.9 with special emphasis on page 239.

181 The most numerous neutral objects in the universe appear to be neutrons, neutrinos and photons. A neutron, however, decays into two charged particles -- a proton and an electron -- and the various families of neutrinos and the photon are never at rest. Accordingly, the ancient approach -- reasonably articulated by M. Born, <u>Einstein's Theory of Relativity</u>, Dover Publications, Inc. (1965) at Section 7., beginning on page 118 -- that assumes the universe to be a void in which neutral celestial objects, neutral clouds of gas and neutral radiation all move can no longer be

in our current conception of the universe and every "mechanical" experimental apparatus we might envision is a participant in the universe's
elaborate electrodynamics. No apparatus to measure time or space
can be developed with operations above the interaction with its sur-
roundings and every charged elementary component of the apparatus
can move only in ways permitted by Maxwell's equations. Further, the
movement of these charged elementary components inevitably causes
changes in the rest of the universe and these changes are communicated at the speed of light by the induction properties of electric and
magnetic fields.

Although Einstein never addressed these matters, substantial work
on the task the author now undertakes has already been done although
almost entirely as an exercise in the understanding of Special Relativity
and without any recognition that what has been done must be considered in assessing gravitational phenomena as well. Thus, a variety of
texts on electrostatics and electrodynamics reach the conclusion that
there is an energy content associated with the existence of an electric field in any given volume of space and that this content can be
calculated by assigning energy of $(E^2/8\pi)dv$ to each volume element
of the "universe" where an electric field is located[182]. Similarly, it is
possible to assign an energy content to magnetic fields based on the

---

defended. As noted previously, the real building blocks of the universe
are all charged and charged objects exert an influence that stretches to
infinity. The only portion of such objects that are voids are their *interiors*.
Conceptually, then, charged objects may more properly be considered
as bubbles in a fluid than shapes in a void. The old conception of the
universe as a spacial vacuum, then, must be discarded.

182 See again E. Purcell, Electricity and Magnetism, (McGraw-Hill Book
Company 1965) beginning at page 51 again with particular emphasis on
the statement: "[t]he potential energy of a system of charges, which is
the total work required to assemble the system, can be calculated from
the electric field itself simply by assigning an amount of energy $(E^2/8\pi)dv$
to every volume element dv and integrating over all space where there
is electric field". The author has referenced this statement several times

square of the intensity of magnetic induction as well[183]. A great deal has been said in many texts discounting the physical significance of these energy assignments and most if not all decline to accept this significance.[184] To the author, however, the only way to make sense of gravitation is to accept the energy as resident in the fields. Otherwise, the gravitating inventory of the universe would be constantly increasing and decreasing as the field energy is absorbed by a discrete object and then re-transmitted back to the field.

Continuing on, then, many texts on electrostatics and electrodynamics also indicate that the existence of a quantized charged source in a rest frame implies the existence of a spherically symmetrical electric field in that rest frame which extends indefinitely and with the field components at a given distance from the "source" particle (as measured in this rest frame) described by Coulomb's law[185]. Applying the concept in the previous paragraph, one can therefore provisionally assign an energy (and, therefore, a mass) quantity of $(E^2/8\pi)dv$ to each volume element of an infinite Euclidian "universe" centered at

before but believes that Mr. Purcell and all who have read his work fail to grasp the full significance of what he has written.

183 See E. Purcell, Electricity and Magnetism, (McGraw-Hill Book Company 1965) beginning at page 256 and L. Page and N. I. Adams, Principles of Electricity, D. Van Nostrand Co., Inc. (1958) at section 46, beginning on page 127 and section 134, beginning on page 481.

184 Note the discussion in Section 5.e., above and especially the portion leading up to the reference in footnote 46. See also, W.K.H. Panofsky and M. Phillips, Classical Electricity and Magnetism, Addison-Wesley Printing Company, Inc. (1956), Chapter 6, especially at the preamble of the Chapter on page 86 and §6-1 on pages 86 to 89 and L. Page and N. I. Adams, Principles of Electricity, D. Van Nostrand Co., Inc. (1958) at section 21, beginning on page 59 with special emphasis on the language on page 61 and section 46 beginning on page 127 with special emphasis on the language on pages 128 to 129.

185 See E. Purcell, Electricity and Magnetism, Section 5.6, pages 158-163 (McGraw-Hill Book Company 1965) beginning at page 160.

the center of the "particle." The unique energy and therefore mass content of each volume element of this field can be translated into an almost infinite number of Euclidian reference frames moving with an almost infinite number of different relative velocities. Because of the transformation properties of electric fields under the Lorentz Transformation equations, the mass/energy content in each of these reference frames will be different and a field that is exclusively an electric field in the rest reference frame will have magnetic components in all others.

All this discussion notwithstanding, the idea that the "particle" is the "source" of the field is a relic of 18th, 19th and 20th century thought. As perhaps no one since Faraday and Maxwell seem to fully and really grasp, it is the field that has existence and content. Thus, if this "source" particle were suddenly accelerated, the former universe of this "particle" as measured in its current and former rest frame would be divided into two distinct regions -- the region to which the change in position of the charge has been communicated by induced changes in its electric and magnetic field components and the region that still perceives the charge as at rest. There is also, of course, a third, transition region in which the details of the process that resulted in the acceleration of the charge is evident.[186] Regardless, however, the "source" particle cannot be the cause of the field -- in our example, the field in most of the universe is inconsistent with the field of its supposed source. Accordingly, the "source" particle is merely partial evidence of what the universal fields look like rather than a true field source. Indeed, the fact that pairs of charged particles of opposite sign can be created and destroyed -- an electron and a positron can be created from an uncharged gamma ray and can extinguish themselves with the emission of gamma radiation -- makes clear that the

---

186 See E. Purcell, Electricity and Magnetism, Section 5.7, pages 163-167 (McGraw-Hill Book Company 1965).

sources do no create the fields but the fields simply cease at their apparent sources[187].

The next element in the understanding of gravitation is to assess the impact of two such elementary charged "point" "particles" on each

---

187 Based on this fact, strictly speaking, the only single quantity that is inevitably and absolutely conserved throughout all aspects of physics is energy. Since conservation of energy and conservation of momentum -- which, with the insights provided by Special Relativity, can be seen to be derivative of energy conservation -- are the cornerstones of the mechanics of Special Relativity, the author suggests that electric field energy identified with a scalar potential field and with a zero set at infinity (usually identified by the character ϕ) be henceforth considered as the most primitive of all electrostatic concepts in physics. The Electric Field intensity, $\mathbf{E}$ (which is the gradient of ϕ) and the source charge density ρ (which is the divergence of the gradient of ϕ) would then be considered derived concepts. So too would the magnetic vector potential (usually identified by the character $\mathbf{A}$) be considered the most primitive of all electrodynamic concepts. The Magnetic Field intensity, $\mathbf{B}$ (which is the curl of $\mathbf{A}$) and the current density $\mathbf{j}$ (which is the curl of the curl of $\mathbf{A}$) would be derived concepts as well. This certainly is not the historical ordering of development, which starts with charge and charge density, then derives the concepts of an electric field and an Electric Field intensity, then develops the concepts of energy and potential, then moves to the concept of a current as a moving charge density, then derives the concepts of a magnetic field and Magnetic Field intensity and finally develops the concept of the magnetic vector potential. In the author's view, however, it is time to impose a rational rather than historical organization on the presentation of eour understanding of electromagnetic phenomena. The present, unthinking reliance on a historical presentation has likely substantially contributed to the current problems physics faces. Further, the concepts of conductors, dielectrics, materials with magnetic susceptibility and such are really not drawn from electrostatics or electrodynamics but, rather, from materials science. For clarity of presentation, it would be well not to introduce these concepts until after vacuum electrostatics and vacuum electrodynamics have been fully developed and it would be better still to henceforth clearly label which concepts are drawn from electricity and magnetism and which are drawn from the study of actual materials with electric and magnetic properties.

other. Fortunately, the consequences of such interactions have also already been developed in detail. A variety of texts have considered first the problem of what a single charged "point" "particle" would experience due to the electric and magnetic field components of a second charged "point" "particle" in motion with respect to it and, second, the corresponding problem of understanding the experiences of the second particle due to the electric and magnetic field components of the first[188].

---

188 The author has placed the word "point" in quotation marks because he does not think that the concept of a "point" source of an electric field has utility. The following discussion -- from E. Purcell, <u>Electricity and Magnetism</u>, (McGraw-Hill Book Company 1965) ¶1.8 beginning at page 20 with special emphasis on the language at the bottom of page 21 -- provides the proper analysis.

> In the neighborhood of a true point charge the electric field grows infinite like $1/r^2$ as we approach the point. It makes no sense to talk about the field *at* the point charge. As our ultimate physical sources of field are not, we believe, infinite concentrations of charge in zero volume but instead finite structures we simply ignore the mathematical singularities implied by our point-charge language and rule out of bounds the interior of our elementary sources. It may be worth noting, nevertheless, that a continuous charge distribution contains no threat of a singularity, and allows the field to be defined at points inside the distribution itself. That is because the volume integral is prevented from blowing up in the neighborhood of $r = 0$ by the fact that the volume element itself goes as $r^2 \, dr$. That is to say, so long as [the scalar charge-density function] remains finite, the field will remain finite everywhere, even in the interior or on the boundary of a charge distribution.

As detailed in a footnote above, this approach, coupled with energy conservation, avoids the infinities that have plagued physics since at least the time of Newton and provides a long required -- but surprisingly rarely sought -- explanation for the stability of fundamental electric moieties. This approach also suggests why quantum mechanics both encounters infinities and is able to resolve them by a renormalization process. The renormalization process simply converts the unphysical conceptual point objects back to their finite charge distributions.

Thus, in W.G.V. Rosser, <u>An Introduction to the Theory of Relativity</u>, Butterworth & Co. (Publishers) LTD. (1964) at section 7.4 beginning on page 290, it is noted that the electric and magnetic field components of the second charged "point" "particle" as measured in the rest frame of the first are determined by what the first particle would compute to be the "current" location of the second. Similarly, the electric and magnetic field components of the first as measured in the rest frame of the second are determined by what the second particle would compute to be the "current" position of the first. This interesting fact bears further comment. The invariance of the field equations and other properties of electrodynamic systems require that the electric field measured at a particular point in space due to a defined uniformly moving or stationary source charge is directed precisely toward the point where the source particle is presently located as measured in the rest frame of the distant interacting particle. It therefore follows that the energy content of the field of this uniformly moving charge at all points in the rest frame space of the interacting particle is due to the present position of the moving charge. The influence of this moving charge at distant points therefore appears to have been transmitted instantaneously although there is no instantaneous transmission of anything. The fact that the motion of the distant source charge is uniform means that there has been time for its distant field to be synchronized with its current position[189].

Continuing on, then, Mr. Rosser's work, at section 7.5 beginning on page 294, provides a further discussion of what is described as a "collision" of two "point" "particles" that are charged. The author believes

---

189 This fact could explain why gravity appears to act instantaneously and thus why the earth responds gravitationally to the "current" position of the sun even though the electrodynamic features of the sun's field take roughly 8 minutes to reach the earth. The electric and magnetic intensities at the earth due to the sun are synchronized to the sun's "current" position.

the following summary and language from that discussion (already included above)[190] is extremely significant:

Mr. Rosser begins by noting that the forces between pairs of moving charges deviate from action and reaction by terms of the order of v^2/c^2. He continues by providing an interpretation of such deviations using classical electromagnetic theory. According to Rosser, citing works by Abraham and Becker and Page and Adams: "[i]t can be shown using classical electromagnetic theory that, if one associates a linear electromagnetic momentum density of $1/c^2$ **E X H** with the electromagnetic field, then the sum of the momentum of the matter and the momentum of the electromagnetic field in an isolated system of charges is constant with respect to time, though the law of action and reaction is no longer true for matter considered on its own". Mr. Rosser illustrates this interpretation with an example that consists of two charges, a charge q_2 moving along a coordinate axis to the right giving rise to fields **E$_2$** and **H$_2$** at a point in empty space and a charge q_3 to the right of q_2 and above the coordinate axis and moving to the left and producing fields **E$_3$** and **H$_3$** at the same point. According to Rosser, [t]he electromagnetic linear momentum per unit volume of this system totals: $1/c^2$ (**E$_2$ X H$_2$** + **E$_2$ X H$_3$** + **E$_3$ X H$_2$** +**E$_3$ X H$_3$**) and "[t]he two terms $1/c^2$ **E$_2$ X H$_2$** and $1/c^2$ **E$_3$ X H$_3$** remain constant, if the velocities of the charges remain constant; [and] these terms are responsible for the electromagnetic masses of the particles when they are accelerated." In contrast, according to Rosser, "[t]he magnitudes of **E$_2$** and **H$_3$** vary at all points in space such that $1/c^2 \int$**E$_2$ X H$_3$** dV, integrated over all of space varies with time and, [s]imilarly, $1/c^2$ $\int$**E$_3$ X H$_2$** dV varies with time. Rosser concludes by noting that work by Page and Adams shows that the quantity "d/dt $\{\int$(**E$_2$ X H$_3$** + **E$_3$ X H$_2$**) $dV\}$ balances the differences between the forces between the charges illustrating how, according to classical electromagnetic theory, the sum of the linear momenta of matter plus the electromagnetic field is constant".

---

190 See Section 5.e., at footnote 45

After we have first defined the field of a single charged particle as we have done and then addressed the way that another charged particle will interact with the field of this first one (and visa versa) we have to address how these two particles will each interact with a third. Here again, well-tested concepts of electro-dynamics provide an answer. We simply have to superimpose the fields of each of these three different particles onto a single frame to get an answer. Nothing in the discussion above has challenged the concept of electrodynamic superposition. Of course, we really have not yet produced a single coherent picture of the fields of these three sources. The measurement of the universal fields of these sources will depend on the frame of reference used and the potentially valid referenced frames are almost infinite. There is, for example, at least one and perhaps as many as three different frames of reference -- one for each source -- in which at least one particle will be at rest. If the three particles we are analyzing represent less than the entire contents of the universe, of course, then no local analysis of the three potential rest frames would allow us to fix which of the three was better than the other two or, indeed, if any of the three is "best".

If these three particles were the only components of the universe, however, there would be a unique frame in which the energy content of the electric and magnetic fields generated by the three particles would be minimized -- there would be a "center-of-mass/energy frame[191]. We could translate all measurements in other frames to this frame using the Lorentz Transformation equations and have at least a start toward discovering a stable and unchanging vantage point from which to assess the behavior of the electric and magnetic fields throughout the universe. This could only be a start, however, because, to find the universal center-of-mass/ energy, we would also have to know the entire history of the motions of the charges as well as their current locations. The energy tied up in the fields induced by this history of changes would have to be considered since the quotation from Mr. Rosser's book above makes clear that there

---

191 See Rindler, W., Essential Relativity, at Section 5.7 (Springer-Verlag 1977)

is a momentum and energy content to the interaction of the fields once separated from the sources and it is this energy and momentum content plus the energy and momentum content of the self-interaction of the field sources that altogether represent the total conserved quantities[192].

In addition to knowledge of the history of the motions of these particles and the current locations of them, we would also have to know the locations of all of the structures in the universe that consist of materials that can be polarized or that are conductors or that have magnetic properties. Such structures would interact with the fields of the sources and the propagating radiation fields, and, in doing so, would cause induced changes to the source and radiation fields. Thus, to understand the universe, we need not only to know the locations of all sources and the rules that dictate changes but also the initial locations of all systems that will respond to the induction properties of electric and magnetic fields.

The author believes the method above ultimately leads to the proper answer although the complexity of the mathematics involved, given the number of components of the universe, likely renders it impossible to translate such an approach into a meaningful large-scale predictive tool[193]. The end result of this approach is that all energy and all mass

192 Which, of course, begs the question whether the energy associated with the radiation acts gravitationally when it is tied up in the field rather than in the localized field sources. If yes, then the reader has already essentially adopted the author's approach of attributing gravitational influence to fields rather than field sources. If no, how is conservation of energy to be maintained? One could cause an object to move when the energy is tied up in the field and then reverse the motion when the energy is associated with the object and thereby obtain more or less gravitational energy from the reverse motion than was required to cause the original motion. Indeed, how would one deal with the analysis provided in footnote 44, above regarding the declining mass of the sun and the increasing mass of its historical radiation under the current gravitational approach?

193 Although physics has been busy attempting to assess the influence of hypothetical "black" "holes" on all manner of things, the author has yet

(and thus all concepts derivative of energy and mass such as force, velocity, acceleration, etc.) can ultimately be tied to the principles of electrodynamics[194].

Thus, as suggested in a footnote above, we can begin with an electro-static scalar potential field representing the electric field energy and with its zero-point at infinity (which typically is identified with the character ϕ and is a scalar function of position) and derive the concepts of the Electric Field intensity, **E** (typically identified as $\nabla\phi$ and a vector function of position) and the source charge density ρ (typically identified as $\nabla^2\phi$ and again a scalar function of position). To this we would need to add a magnetic vector potential (which typically is identified with the character **A** and is a vector function of position) representing the magnetic field energy and derive the concepts of a Magnetic Field intensity, **B** (typically identified as $\nabla \times$ **A** and a vector function of position) and the current density j (typically identified as $\nabla \times \nabla \times$ **A** and a vector function of position). We would further need to link these two distinct energy measurements in an appropriate and fixed way. The author believes that the way to do so is to rely on the equation of continuity $\nabla \cdot j - d\rho/dt = 0$ and impose what has been termed the "Lorentz

---

to see an attempt to model the behavior of a simple Hydrogen atom and a charged particle such as an electron moving toward that atom. Drawing on a variety of different analyses, the author would suggest that the field of the electron would induce a polarization in the different fields of the proton and electron that make up that atom with the result that the charged electron would attract the neutral atom. A rigorous analysis of such an interaction in the same vein as Mr. Rosser's analysis of the interaction of the fields of two charged "point" particles would be extremely interesting and an analysis that considered the mutual inductive effects from a single center-of-mass frame of reference would be best.

194 See L Brillouin, Relativity Reexamined, Academic Press (1970), Chapter 2, Sections 3 to 7 at pages 16-25, for an interesting discussion that largely anticipates the author's approach.

condition" or $\nabla \cdot \mathbf{A} + \mu\varepsilon\, \partial\phi/\partial t + \mu\varepsilon\phi + 0$ on the divergence of $\mathbf{A}$.[195] The use of this condition insures that there is complete symmetry between the scalar and vector potentials, makes both potentials satisfy the same wave equations as those obeyed by the fields and insures a relativistic covariant relation between the scalar and vector potentials. See W.K.H. Panofsky and M. Phillips, Classical Electricity and Magnetism, Addison-Wesley Printing Company, Inc. (1956), Chapter 13 beginning on page 210[196]. Insuring a relativistic covariant relationship between electric and magnetic measurements in local reference frames in relative motion, of course, is the true and only requirement dictated by experiments such as the Michaelson/Morley experiment.

The literature, of course, properly indicates that, if one starts with the typically measured quantities of the Electric Field intensity ($\mathbf{E}$) and Magnetic Field intensity ($\mathbf{B}$) and develops the related potentials by integration, one would have to add integration constants in the form of the time derivative of a scalar field and the gradient of that scalar field, respectively[197]. Further, it seems possible to overlay all manner

---

195 See W.K.H. Panofsky and M. Phillips, Classical Electricity and Magnetism, Addison-Wesley Printing Company, Inc. (1956), Chapter 13 beginning on page 210.

196 It may, indeed, be possible to use the Hertz vector as a more concise basic formulation. The Hertz vector, normally identified with the character Π, has See W.K.H. Panofsky and M. Phillips, Classical Electricity and Magnetism, Addison-Wesley Printing Company, Inc. (1956), Chapter 13 beginning on page 210. Of course, when the material is complex, being too concise may be more of vice than a virtue.

197 Generally, the measurable quantities are the magnetic intensity ($\mathbf{B}$) which is the Curl of the Vector Potential $\mathbf{A}$, and the Electric Field intensity ($\mathbf{E}$), which is the gradient of ϕ. As a result, the measurable magnetic intensity would be the same whether the Vector Potential is defined as $\mathbf{A}$ or a quantity $\mathbf{A}'$ that satisfies the equation:

$$\mathbf{A}' = \mathbf{A} - \text{grad } \Psi$$

of additional vector fields atop the electromagnetic vector fields in the same physical space.

Note, however, that, if one starts with the potentials and not field intensities, the differentiation of potentials does not require the addition of constants. Moreover, while it certainly is possible for many vector fields to co-exist in the same physical space, it is not possible for the source density of the electric and magnetic fields to respond to attributes of the other fields -- such as an overlaid gravitational field with its own set of energy rules -- if the overall requirement of conservation of mass/energy is to be preserved for even one of the fields.

To illustrate, consider a single electron left free in space in the presence of a gravitational "field" as such a field is currently conceptualized. Existing thought (and, of course, substantial experimental proof as well) indicates that this electron will be accelerated by that "field". The acceleration of this electron is evidenced by changes in the Electric Field and Magnetic Field intensities of which this electron is purported to be a source as standard texts on electrostatics and electrodynamics would confirm. These changes will propagate from the point location of the accelerating electron at the speed of light, and, based on the discussion above, these changes all can only occur at a localized energy cost[198]. To preserve the concept of conservation of energy and

---

Similarly, the Electric Field intensity would be the same whether the Electric Potential is defined as ϕ or a quantity ϕ' that satisfies the equation:

$$\phi' = \phi - \partial\Psi/dt$$

The transformations defined by these two equations are known as gauge transformations. See W.G.V. Rosser, An Introduction to the Theory of Relativity, Butterworth & Co. (Publishers) LTD. (1964) at Section 9.1 beginning on page 348 and especially at 350 to 351.

198 As the reader contemplates the acceleration of this electron by the gravitational field toward a "mass" center, compare this analysis with the analysis of the collapse of a gravitating body (including conceptually, an

avoid instantaneous transmission of that energy, the energy cost must be charged against the space in which the increased electrical and magnetic energy are now located. For this to be true, the increases in the electrical and magnetic energy in the field of the formerly stationary electron must be precisely matched by decreases in the energy of what has traditionally been described as the gravitational "field". The energy of the gravitational "field", then, can only be electromagnetic in origin and, thus, gravity must be a property of electromagnetic systems[199].

It is true, of course, as physicists of the late 19th and 20th centuries discovered -- and, indeed, became transfixed by -- that there are an infinite number of Euclidian frames of reference in which the vector and scalar fields identified above can be defined and manipulated. Further, there are no purely local experiments by which any one of these frames may be determined to be any better than any of the others. There is, however, no reason why anyone should rely on purely local experiences in order to make such a determination. Physics stuck in an elevator simply will not do. Instead, physicists can use the navigational

---

electron) into a "black" "hole" as discussed in subsection b of section 11, above. In the current analysis, the accelerating electron radiates away the energy it accumulates as it falls. In the analysis of a "black" "hole," the accumulating energy of the falling "object" would need to stay associated with the object so that it can be carried into the "black" "hole." Here again, the author's view that there are no "neutral" objects in the universe indicates that the former analysis is accurate and the latter is fiction.

199 The fact that there are gravitational red and blue shifts of electromagnetic radiation should only serve to confirm this. As far as the author can determine, there is universal agreement that electromagnetic radiation consists of nothing other than mutually perpendicular Electric Field intensities and Magnetic Field intensities propagating in the direction of their common normal. The propagation along this normal is due to the mutual inductive properties of these fields. Accordingly, as the field intensities decline due to gravitation in moving through a element of space, all of the electromagnetic field energy lost as a result must be matched by an increase in the energy of the gravitational "field." Thus again, the energy stored in a gravitational "field" must be electromagnetic in origin.

and astronomical skills they have developed in the past several millennia to establish the center-of-mass frame of the Solar System as a rough stand-in for this universal center-of-mass. As further work is done, what is discovered in reliance on this frame can be translated into the more complete center-of-mass frames of the future.

The approach above[200] suggests two candidates for processes that, either alone or acting in concert, explain the mutual attraction of all electrodynamic structures that have rest mass. Gravitation may arise from the fact that any observer who accelerates with respect to the center-of-mass frame of the bulk of near-by electrodynamic structures absolutely will not find his or her frame equally suitable for the description of the evolving electromagnetic structures he or she experiences. The now moving observer (or the now moving train in which the observer is standing and at risk of falling) will suffer apparent length contractions, see time dilations, measure contracted electric source densities and measure electric current densities that are suddenly different as comparted to the surrounding bulk material. The contraction, dilation and other electrodynamic effects will render the perspective of this observer disadvantaged as compared to an observer at the center-of-mass of the surrounding bulk matter. Further there are elaborate interrelations between electric charge, electric polarization and magnetic effects in bulk matter that will cause complex mutual interactions between the electrodynamic structures that make up the bulk matter under observation when those structures are viewed from the preferred and the disadvantaged frames. These may very well explain gravitation. Indeed, just as magnetism is a facet of electrodynamics when viewed from different uniformly moving reference frames, gravitation

---

200 The author, of course, must acknowledged that the approach above he is referring to is hardly an original one (and, indeed, that there is very little in this work that is truly original other than the rearrangement of unoriginal ideas in a very original way). Thus, the reader can see a similar approach outlined in J.C. Slater and N.H. Frank, Introduction to Theoretical Physics, McGraw-Hill Book Company, Inc. 1933) in Chapter XXI with special emphasis on Section 153 beginning at page 241.

may similarly be a facet of electrodynamics viewed from changing reference frames[201].

Further, one critical aspect of electrodynamics that typically is ignored in most expositions of its features is the continuous interaction of each charge density element with all of the other charge density elements in the universe. Thus, while the illustration in Mr. Rosser's work of the motion of two charged objects moving past each other at a constant velocity is interesting, it is also artificial. The mutual induction affects of the fields of these two particles greatly complicates a rigorous analysis.

A natural response to the author's view that the universe is purely electrodynamic in nature is to note the well-established facts that there are some forms of mass/energy that are "mechanical", "thermal", "chemical" and, thus, not traditionally assigned as electromagnetic. Moreover, the ability to concentrate electrically charged things of the same sign would be impossible if all forces were purely electrical in

---

201 It is now clear that it is more than a coincidence that the attractive force between two permanent magnets obeys an inverse square law just as the force between two stationary charges obeys an inverse square law. It is, in fact, the very same law and, indeed, the very same force. A variety of texts on electrodynamics indicate that the transformation properties of steady currents – like the steady atomic currents in a permanent magnet -- cause some objects viewed as neutral in one unique reference frame to nevertheless be charged in all other relatively moving frames. See E. Purcell, Electricity and Magnetism, (McGraw-Hill Book Company 1965) at ¶5.9 beginning at page 172. Accordingly, the force between magnets falls off as $1/r^2$ because, in reality, it is the very same force as between static charges save that its operation only becomes apparent due to the relative motion of the two frames. The Lorentz Transformations required to be used by different observers in relative motion insure that electrical neutrality is frame specific and a first order effect of the Lorentz Transformation equations required to analyze objects from relatively moving reference frames is magnetism. Gravitation may simply be a second order effect of the same transformation equations.

origin -- the concentration of such things must work against electro-static repulsion and therefore cannot be caused by a purely "electric" field. Further, some writers on electrodynamics have suggested that the self-energy of even a simple electron is inadequate to explain its inertial behavior[202]. We will see, however, as detailed below, that all of the energy forms referenced above ultimately, when viewed at the atomic level, involve the behavior of electrically charged things and only such things. Further, although we cannot use a pure electric field to concentrate identically charged things, we can, as discussed below, manipulate combined electromagnetic fields to do so. Finally, serious effort has been undertaken to attribute all of the inertial behavior of all electrodynamic moieties to induction effects similar to those embodied in Lenz's law and the author believes that those efforts, when erroneous concepts are purged, will ultimately explain all of the behavior of electro-dynamic systems, including the behavior we now term as gravitational.

Beginning then with arguments in support of an electromagnetic character for all forms of energy, the author starts by analyzing the energy and momentum carried by electromagnetic radiation. Most advanced texts on electrodynamics provide discussions of such energy and momentum transfer. Thus, for the energy carried by radiation, Page and Adams[203] indicate:

> In the general case, however, the law of conservation of energy reads: the rate at which the electromagnetic field does work on the charges in the volume τ plus the rate at which energy flows out in the form of radiation through the surface S bounding τ must equal the rate of decrease of electromagnetic energy in this region. Thus, we conclude that the second term on the left

202 See, M. Born, Einstein's Theory of Relativity, Dover Publications, Inc. (1965) at 210 to 212 and later at 286 to 287.

203 L. Page and N. I. Adams, Principles of Electricity, D. Van Nostrand Co., Inc. (1958) at section 134, beginning on page 481 with the text quoted drawn from pages 483-484.

of [the equation $m_\tau \mathbf{j} \cdot \mathbf{E} \, d\tau + m_\sigma \boldsymbol{\sigma} \cdot d\mathbf{S} = \partial/\partial t \, m_\tau (\kappa\kappa_0/2 \, E^2 + \mu\mu_0/2 \, H^2) d\tau]$ represents the energy passing through the surface S per unit time in the form of radiation. From the form of this term it is clear that we can assign to each unit area of cross section a rate of flow of energy given by the vector

$$\boldsymbol{\sigma} = (\mathbf{E} \times \mathbf{H})$$

which is known as the *Poynting flux*. Its direction is that of the flow of energy and its magnitude specifies the quantity of energy passing through a unit cross section in a unit time. It is to be noticed that $\boldsymbol{\sigma}$ vanishes unless both electric and magnetic fields are present and not in the same direction. The flux of energy is perpendicular to the plane of $\mathbf{E}$ and $\mathbf{H}$ in the sense of advance of a right-handed screw rotated from the first to the second of the two vectors. It is proportional in magnitude to E, H and the sine of the angle between them.

As for the momentum component, Page and Adams[204] indicate:

Consider next a region τ containing radiation but no ρ or $\mathbf{j}$. The force $F\tau$, then, must be zero. Therefore, if $F''\tau$, is the force on this region due to the stresses acting over the surface,

$$0 = F'' - \partial/\partial t \int \tau \, (\kappa\mu/c^2 \, \boldsymbol{\sigma}) \, d\tau$$

As the volume τ contains radiation and nothing else, the force F'' due to the stresses on its surface must be supposed to act on the radiation in its interior. As we have

$$F'' = \partial/\partial t \int \tau \, (\kappa\mu/c^2 \, \boldsymbol{\sigma}) \, d\tau$$

---

204 L. Page and N. I. Adams, <u>Principles of Electricity</u>, D. Van Nostrand Co., Inc. (1958) at section 138, beginning on page 494 with the text quoted drawn from pages 496-497.

and as the force is equal to time rate of increase of momentum, the radiation has an *electromagnetic momentum*

$$\mathbf{G} = \int \tau \, (\kappa\mu/c^2 \; \boldsymbol{\sigma}) \; d\tau$$

We may attribute momentum, therefore, in the amount

$$\mathbf{g} = \kappa\mu/c^2 \; \boldsymbol{\sigma}$$

to each unit volume of a radiation field. For plane waves, the electromagnetic momentum per unit volume has the magnitude[205]

$$g = \kappa\mu/c^2 \; vu = u/v$$

Similar discussions found in Scott[206] and Rosser[207] confirm general agreement among physicists that electromagnetic radiation carries both energy and momentum in calculable quantities. Not surprisingly, then, physicists have contemplated the use of electromagnetic radiation to power a rocket -- the so called photon rocket[208]. The obvious

---

205 In this equation, v is the phase velocity of the radiation and u is the total energy density. The second identity makes use of the equation $v=c/(\kappa\mu)^{1/2}$

206 W.T. Scott, The Physics of Electricity and Magnetism, John Wiley & Sons, Inc. (1966) at section 10.4, beginning on page 553 with the discussion of the energy content found on pages 554-557 and the momentum discussion found on pages 557-559.

207 W.G.V. Rosser, An Introduction to the Theory of Relativity, Butterworth & Co. (Publishers) LTD. (1964) at Section 5.8.3 beginning on page 222 and Section 9.5 beginning on page 363.

208 Physicists have also contemplated the use of a "solar sail" for similar purposes. The author, therefore, likely could use a distant source of electromagnetic radiation and a perfect reflector as the basis for the argument that is to follow. The author has not done so because the texts he has reviewed focus extensively on the momentum transfer to a perfect reflector (at twice the momentum of the incident, reflected radiation)

benefit of such a rocket is that it has the optimum exhaust velocity -- because no tangible thing may move faster than the speed of light, no rocket ejecting an object, combustion product or other thing with a rest mass can equal the exhaust velocity of a photon rocket[209].

The fact that physicists have contemplated the use of such a rocket makes clear that the energy and momentum in electromagnetic radiation may be converted into perhaps the most basic form of "mechanical" energy -- kinetic energy. The author finds this significant because a "photon" rocket contemplates an ostensibly "neutral" phenomena, electromagnetic radiation, causing the motion of an ostensibly neutral object, the rocket itself, through energy and momentum relations that are linked to the behavior of electromagnetic fields and nothing else. Indeed, the arguments in the texts cited above providing the foundation for the assignment of energy and momentum to electromagnetic radiation go further. These indicate that the motion of the exhaust radiation causes motions of the charged particles that make up the rocket and changes to the fields of interaction of these particles so that the momentum of the exhaust radiation and of the rocket are precisely matched. To confirm that this is true, consider the following discussion from the text by Scott[210] referenced in a footnote above:

> but don't spend much time on energy conservation. If energy is to be conserved, the reflecting object could not move unless the reflected radiation suffers an energy loss equal to the kinetic energy gained by the reflecting object in the momentum transfer. The explanations the author has seen suggest that the reflected radiation would be shifted to the red as it is re-radiated from the conductor surface because of the Doppler effect and that the energy and momentum relations can be balanced. The author has no doubt that this is true -- otherwise we would have found a perpetual motion machine -- but prefers to avoid plowing new ground and thus relies on the photon rocket, instead.

209 See French, A.P., Special Relativity, W.W. Norton & Company, Inc. (1968), pages 183-184.

210 W.T. Scott, The Physics of Electricity and Magnetism, John Wiley & Sons, Inc. (1966) at section 10.4, with the quotation drawn from page 555.

We find for the rate at which work is done on all the charges in an arbitrary volume τ the expression:

$$P = \int_\tau P_\tau \, d\tau \; = - \int_{\text{surface of }\tau} \mathbf{E} \times \mathbf{B} \cdot \mathbf{dS} - \partial/\partial t \int_\tau (B^2/2\mu_0 + \varepsilon_0 E^2/2) d\tau$$

To interpret this result, we consider first a very large volume whose surface lies outside the region in which the fields of interest have any appreciable value. Then the surface integral vanishes, and we see that the rate of doing work is equal to the rate of decrease of a certain volume integral which can be interpreted as the energy contained in the electromagnetic field:

$$U = \int_\tau (B^2/2\mu_0 + \varepsilon_0 E^2/2) d\tau$$

which is often written in symmetrical fashion as

$$U = \int_\tau \tfrac{1}{2}(\mathbf{B} \cdot \mathbf{H} + \mathbf{D} \cdot \mathbf{E}) d\tau$$

This latter equation remains valid for linear dielectric and magnetic materials if we do not count polarization charges and magnetization currents in ρ and $\mathbf{J}$ but include them by using the $\mathbf{D}$ and $\mathbf{H}$ vectors (Problem 10.4e). In this case, P_τ represents only the power expected on free charges, the work done in changing polarizations and magnetizations being included in U. For nonlinear media, hysteresis and dielectric losses must be included, leading to complexities beyond the scope of this book.

*   *   *

Now if we consider a smaller volume so that the surface integral does not vanish, we can find an interpretation for it as well. Using again symmetrical notation that holds in general, we have

$$\int_S \mathbf{E} \times \mathbf{H} \cdot \mathbf{dS} = -P - dU/dt$$

so that the integral is equal to the rate of gain of energy from the charges -P, plus the rate of decrease of the field energy within the surface. That is, if conservation of energy is to be maintained, the surface integral must represent the rate of outflow of energy through the surface. We can then assign to each unit area of the surface a rate of outflow energy given by

$$N = E \times H$$

where **N** is the called the "Poynting vector," happily named after the physicist J.H. Poynting.

If we simply chose an outflow surface that is the instantaneously boundary between the exhaust and the body of the rocket (and unspent fuel), these relations mean that the energy and momentum of the exhaust is obtained by extracting energy from and providing equal and opposite momentum to the charges and the fields on the rocket side of the surface. The kinetic energy and momentum of the rocket are, therefore, associated with the motions of the charges and fields that make up the rocket. If we assume that the rocket consists of nothing but charges and their fields -- if we accept the atomic theory of matter as the author believes virtually all, if not, indeed, all scientists do -- then that should be the end of the discussion[211]. Kinetic energy is exclusively an electromagnetic phenomena.

---

211 Indeed, because the energy and momentum of the exhaust is exclusively electromagnetic in conception, how can the energy and momentum of the "thing" that is now moving in the equal and opposite direction as a result of the energy and momentum transferred from the exhaust be anything else? Further, if the energy and momentum of the exhaust is electromagnetic in conception and virtually all of the substance of the "rocket" will ultimately be converted to electromagnetic exhaust radiation, we face the paradox that, as the rocket and payload become a smaller and smaller portion of our system, this smaller and smaller portion would have to have more and more of its mass and energy tied up in the something that is other than the electromagnetic arrangement of its parts.

Once a neutral object has been set in motion with the use of electro-magnetic energy, this coherent kinetic energy can easily be converted into another ancient energy form, heat. All that must be done is to position some object into the path of the rocket that will collide inelastically with it. In the resulting inelastic collision, some of the coherent energy of motion formerly existing will be transformed into the incoherent energy of motion traditionally labeled as "heat". Since the energy of the original motion was tied to the charged particles and fields that make up the rocket, the energy we describe as heat must also be associated with the energy and fields that make up the new object into which the rocket has been incorporated. Indeed, to confirm that heat is an electromagnetic phenomena, the author notes the following discussion from Scott[212]:

> Another demonstration that we cannot picture energy as being carried by a current is found if we consider the [direct current] case of a length L of a long straight cylindrical wire of resistance R, and of radius r. If the current i flows in it, the [potential difference] between the ends will be $V = iR$, and the electric field component parallel to the wire at its outside will be $E_x = V/L = V = iR/L$. The magnetic field H at the surface of the wire will be $H_\varphi = i/2\pi r$. The Poynting vector $\mathbf{N}$ will then be directed into the wire, as a little consideration will show, and will have the magnitude
>
> $$N = | \mathbf{E} \times \mathbf{H} | = E_x H_\varphi = i^2 R/2\pi r L$$
>
> Since $2\pi r L$ is the area of the outside of the wire, we see that the rate of flow of energy through its sides is
>
> $$\int \mathbf{E} \times \mathbf{H} \cdot d\mathbf{S} = 2\pi r L E_x H_\varphi = i^2 R$$
>
> which is the rate at which heat is developed. That is, all of the Joule heat enters the wire through its sides in the form of the

212 W.T. Scott, The Physics of Electricity and Magnetism, John Wiley & Sons, Inc. (1966) at section 10.4, with the quotation drawn from page 556.

Poynting flux and none enters through its two ends. This is actually not surprising when we consider that the current involves a continual jumping of electrons between energy levels, without any significant action of one electron on another. **It is the fields that make the charges flow, and it is the fields from which this energy is derived**.

Accordingly, all of the energy of motion of objects, be it coherent or incoherent, is associated with the energy content of the fields of the charged components within the objects and the field attributes that hold these charged components together[213]. The most basic forms of "mechanical" energy, are, therefore, really attributes of the electromagnetic fields and field sources that make up the objects behaving in these "mechanical" ways.

We have not yet, of course, discussed the various forms of potential energy such as the energy stored in chemical bonds or springs and such. The standard description of chemical energy, however, is that energy which can be obtained (or that is consumed) in rearranging the electromagnetic bonds of the reacting chemicals. The standard explanation for how energy is stored in a compressed spring is that there is an equilibrium position of the atoms in the molecules or lattices that make up the spring and that the compression or extension of the electromagnetic bonds between these atoms generates an electromagnetic restoring force[214] which operates to reconvert this "potential" energy

---

213 Indeed, if we can identify even one form of motion such as heat whose transfer is entirely associated with alterations to the positions of the electromagnetic components of a substance, we really have made it impossible for there to be any forms that are not associated with such alterations. Otherwise we could apply the non-electromagnetic heat form to a substance that we could move electromagnetically and produce field changes that are inconsistent with Maxwell's equations.

214 See J. Jewett, Jr. and R. Serway, Physics for Scientists and Engineers, 6th Ed. (Thompson Brooks/Cole 2004), Section 8.6, especially Example 8.11 on pages 237-238.

into kinetic energy of some object in contact with the spring. The force exerted by the spring, therefore, is simply the superposition of these electromagnetic restoring forces.

Indeed, a variety of scientists over the centuries have remarked on the similarity between the mathematical rules that govern the behavior of electricity, magnetism and heat transfer. Further, it has always been somewhat of curiosity that the very same mathematics that apply to a driven harmonic oscillator can be used to describe the behavior of circuits that have capacitance, inductance and resistance. Of course, if the author is correct, the similarity in the behavior of these physical attributes of systems is neither coincidence nor curiosity. The mathematics for one applies to all of the others because the behaviors are all linked to the activities of electrodynamic moieties constrained to follow Maxwell's equations.

Ultimately, then, to bring this argument back to its beginning, the author would note that most of the energy used by mankind inevitably can be traced to the capture of radiation from the sun. The energy of large-scale power generation comes, for the most part, from chemical potential energy released through the combustion of coal, oil, natural gas, wood and other biomass. Forms that rely on such chemical potential energy ultimately rely on the ability of green plants (either recently living or ancient) to capture the energy in solar radiation to convert carbon dioxide and water to complex carbon chain molecules through elaborate chemical reactions. Mankind generates power by reversing these reactions. In the usual scenario, the reverse reactions create heat which produces mechanical energy which, in turn, generates electricity. Regardless of the scenario by which power is generated, however, it is clear that all of the chemical energy resident in fossil and biomass fuels is really radiant solar in origin.

Further, although the mechanism is different, another material source of generated electricity, hydroelectric power, also depends on radiant solar energy. Radiation from the sun causes water to evaporate, creates temperature differences leading to currents of air which transport the water vapor such that, when it condenses and falls as rain, it does so at a place of higher gravitational potential than where it evaporated. Hydroelectric power simply taxes the return of this water to its original low-potential position. Accordingly, virtually all of the power used on earth can be traced to the capture of solar radiation[215].

Given these facts, how could solar radiation act at the atomic level except by the operation of its electric and magnetic energy components? What else is there to such radiation? Further, given that there is nothing other than electric and magnetic components to such radiation, when such radiation is absorbed and its energy and momentum are transferred to the charged components of the absorbing substance as indicated in the various references detailed above, what resulting motions of those charged components are possible other than those absolutely dictated by Maxwell's equations? If the charged components move is some way other than as dictated by Maxwell's equations, then energy and momentum cannot be universally conserved for we could retain the excess non-electro-magnetic energy of motion of these components simply by reversing the absorption process and allowing the substance to radiate.

Of course there is the problem of causing the concentration of electrically charged objects of the same sign. Forcing identically charged

---

215 There are, to be sure, primary terrestrial energy sources that do not depend on the sun. These include the thermal energy resident in the interior of the earth, the energy liberated by decay of the radioactive components of the earth and energy generated by nuclear fission or fusion of components of the earth. As to the former, as discussed previously, the author believes that heat energy is electromagnetic in concept. As to the latter two, it is possible to link the energy of radioactive decay and of nuclear fission and fusion to the energy stored in electromagnetic moieties although the author must concede that he can hardly be said to have made a compelling case on this point.

objects to move closer obviously cannot be accomplished by a pure electric filed. The basics of modern power generation, however, make it obvious that we can generate large pure electrostatic fields through the use of magnetic fields that are changing in time. Modern society produces tremendous voltages (and thus potential energy differences) by moving conductors through the fields of powerful magnets. Indeed, half of the energy content of solar radiation is associated with the magnetic component of the incident radiant energy so half of the historical energy tied up in fossil fuels and biomass is not "electrical" but rather "magnetic" in origin. Of course, the separation of electromagnetic fields into "electric" and "magnetic" components is yet another historical artifact. There is a single field in a single universe that exhibits different properties that we ascribe to "electric" and "magnetic" components as our velocity through this single universe varies.

Finally, we should comment on prior apparent failures to attribute all inertia to electrodynamics. Thus, for example, according to Born, the static energy of an electron is three-quarters electrostatic and must be one-quarter something else[216]. The author has found additional materials through the internet, however, that suggest that Born's approach is mistaken. Thus, according to a paper entitled "The Theory of the Electron" by F. Rohrlich published in 1962 (and available at no cost over the internet):

> The mathematical formulation of *macroscopic* electromagnetic phenomena was beautifully accomplished by J. C. Maxwell about thirty years prior to Lorentz's theory. Lorentz construct a *microscopic* theory by using Maxwell's equations and adding to it an expression for the force which a charged particle experiences in the presence of electric and magnetic fields. This microscopic theory is a description of matter in terms of its charged atomic fragments, ions and electrons. The success of

---

216 See, M. Born, Einstein's Theory of Relativity, Dover Publications, Inc. (1965) at 287.

this microscopic theory lay in the proof first provided by Lorentz that the macroscopic Maxwell theory can be deduced from this microscopic theory by a suitable averaging process over the motion of the individual ions and electrons. Thus, Lorentz's theory became the primary theory and Maxwell's theory can be reduced to it.

However, Lorentz went beyond this: having successfully described the electromagnetic force acting on a charged particle due to externally present fields, he attempted to describe the structure of an individual electron. His aim was to show that the electron is a completely electromagnetic object. In particular, its mass was to be the mass equivalent of its electromagnetic energy contents; its inertia, i.e., the inertial term in Newton's equations of motion, was to be entirely due to its own electromagnetic field. Accelerating the electron means changing or distorting the field produced by the electron; this requires work. Therefore, the electron exhibits a certain inertia in following the force acting on it.

These ideas were contained, partly implicitly, in Lorentz's work. They were later clarified and extended to fast moving electrons by Abraham and others especially after Einstein's special theory of relativity became accepted. However, relativity was not applied to the theory in a consistent way and, consequently, the difficulty that the union of relativity and electron theory could have removed, remained there for many years. Let us therefore look at the non-relativistic theory.

As I mentioned previously, the starting point of the theory is the Lorentz force and the microscopic equations which, when averaged, produced Maxwell's equations. When the Lorentz force is used to describe the action which the electron's own electromagnetic field exerts on its source, the electron, the following equation of motion is obtained:

$$4/3 m\mathbf{a} - 2/3\ e^3/c^3 \mathbf{\mathring{a}} + \text{"structure terms"} = F \quad (1)$$

In this equation the acceleration **a** and the force **F** are three-vectors and the dot indicates a time derivative. [The] m and e are mass and charge of the electron and c is the velocity of light. The famous Lorentz force **F** is due to the external electric field **E** and magnetic flux density **B**, and is the sum over the charge density r,

$$\mathbf{F} = \int \rho(\mathbf{E} + \mathbf{v} \times \mathbf{B})\ d^3x \quad (2)$$

The electron's structure is therefore characterized by a charge density distribution. This quantity has to be assumed, since there is nothing in the theory which would determine it. Only the total charge e is known from experiment.

The mass m is completely electromagnetic and is therefore given by the energy W of the electric field when the particle is at rest,

$$m = W/c^2 \cong e^2/r_0\ 1/c^2 \quad (3)$$

In this equation the radius r_0 of the electron enters.

Equation (1) now exhibits three difficulties, associated respectively with each of the three terms appearing on the left hand side:

(I) The inertial term differs by a factor 4/3 from Newton's classical "mass times acceleration." This is a kinematic problem which implies that the relationship between momentum and velocity for a particle in Newtonian mechanics differs from that for the completely electromagnetic electron. This would have dire consequences and is an intolerable defect in the theory. Luckily, it can be corrected rather easily. A conscientious merger of this theory with special relativity assures that this factor disappears, since it is incompatible with the relativistic transformation properties. For a finite electron this was first pointed out by Fermi in 1922. It is closely related to the definition of rigidity in special relativity where the difference in simultaneity of relatively moving observers plays an essential role. Unfortunately, Fermi's paper was either never understood or soon forgotten. In any case, the factor 4/3 can still be found in some of today's texts. For point electrons, the removal of this factor was later rediscovered several times. Let me summarize by saying that a relativistic generalization of (1) will not show this factor; if the non-relativistic theory is derived as the limit of the relativistic one, this factor will disappear from (1). A study of its origin reveals an unjustified and incorrect assumption about the relationship between the Poynting vector and the momentum of non-radiative electromagnetic fields.

●   ●   ●

Mr. Rohrlich goes on to address problems associated with first the "structure terms" and then the middle term of equation (1) and those that are interested may seek Mr. Rohrlich's paper on the internet. The author, of course, has his own solution to the problems with the "structure terms" and what Mr. Rohrlich's paper, in a portion not reproduced above, describes as the related problem of infinite electron self-energy. That solution is outlined in footnote 21, above. As far as the middle term is concerned, the author would not be surprised to find that this middle term or the manipulations needed to eliminate it ultimately provide the key to understanding gravitational behavior. Moreover, the formulation of the problem by Mr. Rohrlich carries over many prior faulty concepts and gravity may appear once these faulty concepts are purged.

Thus, for example, equation (3) from Mr. Rohrlich's paper treats all of the electron's energy content as being localized near the center of the electron when, in fact, that energy is spread throughout the universe -- although most of that energy can certainly be found very near the electron center[217]. Further, the transformation from classical to relativistic formulations which is described in Mr. Rohrlich's paper as the solution to the "4/3 problem" does not adjust the results of those formulations to a master, center-of-mass, system as the author believes is required.

Regardless, however, the author, sees nothing in the behavior of objects in our everyday universe that absolutely cannot ultimately be linked to electromagnetism. To be sure, we have not yet provided a coherent picture of the atomic nucleus and nothing in this work suggests how the so-called "strong" force between protons and neutrons operates. The charge of particles that participate in the strong force is both identical to that of electrons and conserved, however, so that it would not be impossible to construct a theory of the strong nuclear force which preserves the correspondence between energy stored in atomic nuclei and the energy content of the electromagnetic fields of the nucleons. Certainly, all of the indicia of radioactive decay of atomic nuclei are electromagnetic. Gamma radiation is simply electromagnetic radiation. Beta radiation is simply rapidly moving electrons and alpha radiation is simply Helium

217 As suggested previously, the conceptual model for an electron should no longer be a pebble in a vacuum, but, rather, a bubble in a fluid. Those who apply the former visualization are obviously troubled by the tendency of their model to grow toward an infinite localized mass as the acceleration of their conceptual pebble continues for a longer and longer period of time. The latter visualization, however, suggests the impropriety of imagining long periods of acceleration of the phenomena under consideration. One cannot divorce the behavior of a bubble from the behavior of the fluid and it is, therefore, entirely inappropriate to imagine the bubble as moving faster and faster outside the context of the fluid in which it exists. Indeed, the fact that the mathematical model for an electron does not make sense when accelerating for long periods of time becomes reassuring rather than troubling under the latter approach.

nuclei in rapid motion. Accordingly, when the strong force fails and a system formerly held together by that force disintegrates, all of the disintegration products are electrodynamic moieties whose motion and rest mass define the energy content of the post-disintegration system[218].

Ultimately, then, the author believes that the essence of gravity will be found by taking what is know about electricity and magnetism, adopting a master, center-of-mass frame of reference and then resolving the behavior of "objects" into two distinct categories. First there is the behavior of the fields that pervade all of space and surround and penetrate the "objects" and then there are the complex induced changes to the "objects" unique to the electromagnetic signatures of the components of those objects.

---

218 Also significantly, the precise neutralization of the positive charge of a proton by the negative charge of an electron means that the dilute charge density at infinity of both particles has to be precisely the same. Otherwise, the placement of a proton inside the sphere of an electron would leave a residual field. See E. Purcell, Electricity and Magnetism, (McGraw-Hill Book Company 1965) at section 2.8 beginning at page 51 with particular emphasis on the statement: "[t]he potential energy of a system of charges, which is the total work required to assemble the system, can be calculated from the electric field itself simply by assigning an amount of energy $(E^2/8\pi)dv$ to every volume element dv and integrating over all space where there is electric field".

13. Conclusion

The author hopes that he has brought some clarity to the searches for Dark Matter and Dark Energy. Indeed, he hopes that he has established that, properly understood, these searches are not for any new matter or energy at all. Instead, they are searches for a theory of gravitation that can account for the strength of gravitational "fields" based on the amount of tangible matter (and implicit electromagnetic energy and thus mass) that can be observed.

The author further hopes that, in the process of setting forth his views on the dark features of the universe, he may also have shed real light as well on other issues of concern in modern physics. Thus, for example, the discussion above to some extent "unifies" gravitation and electrodynamics by subordinating the former to the latter and provides a new and simple explanation for the stability of electrodynamic structures.

Perhaps as importantly, however, the author hopes he has demonstrated that much of what is current taken as true in physics is simply wrong -- so very wrong, in fact, that one need not become an expert in the mathematical methods that are currently popular to be confident that physicists long ago lost touch with reality. If the author accomplishes nothing else, then, he hopes he has returned the focus of physicists to reason and logic. Mathematics, though certainly a vital tool to the understanding of the real world, clearly can lead one very far astray from the truth. Indeed, once bad logic is buried under complex mathematics there is no limit to the mischief that may follow[219].

As can be expected, the discussion above leaves many, many issues unresolved. Thus, there is only an outline of how the attraction of objects to each other that we ascribe to gravitation really works. There also is no explanation in this work for why electric charge is quantized. Further, although the discussion above makes the wave properties of fundamental particles and the particulate properties of electromagnetic radiation less mysterious -- the particles and radiation are simply different aspect of one single phenomenon, the electromagnetic field, so all features of this single phenomenon should have such dual natures -- it is not consistent with many concepts developed in quantum electrodynamics. The author, of course, is under the impression that General Relativity and quantum electrodynamics were never reconciled so his approach to gravitation can hardly be said to be radically more out of harmony with quantum electrodynamics than the existing one.

---

219 The author feels compelled to add that, notwithstanding what are undeniably harsh criticisms of the work of existing physicists, he views it important that every one who has carefully considered a problem be allowed to state his or her views without fear of ridicule. Time and again, ideas that arose well outside the mainstream of thought and that originally drew the condemnation of "experts" have come, with the passage of time, to be accepted as substantially correct. The theory of continental drift as the forerunner of plate tectonics is an obvious example. Ultimately, then, the author hopes his undeniable hypocrisy will be forgiven. He has read extensively in the works of modern adherents of General Relativity and developed a firm conviction that those adherents have served mankind very poorly and themselves rather richly. If harsh words is all that they receive as a consequence of their failings, they should consider themselves lucky indeed. There are few areas of human endeavor in which such substantial failure draws so little blood. The author, then, is firmly of the view that the scientific community must listen to every man's views and evidence. Such surely will result in some wasted time yet, in the last 100 years or so, how much time have gravitational physicists spent wisely?

This work also certainly fails to integrate the "strong" nuclear force into either gravitation or electrodynamics and gives no insights into the interior of the atomic nucleus. While the author's intuition is that the mass of nucleons is electrodynamic in origin -- the intense concentrations of electric energy in charged nucleons accounts for their inertia -- this certainly has not been proven.[220] The author frankly has never found a truly accessible discussion of the "strong" force, does not understand it, and must leave the task to others or a much later time.

Despite the limitations of this work, the author hopes that his ideas will assist others in the further development of the most basic and pure of sciences. Even more so, however, he hopes that these others, many of whom have the luxury of devoting their full time and attention to the task, will not forget that their task is not complete until they have made their truths **_easily_** accessible beyond the ivory towers in which they work.

---

220 The discussion above, however, is consistent with the behavior of bulk matter that we have come to expect. Bulk matter, with its massive positive nuclei and much less massive negative electrons, is constantly undergoing induced polarizations caused by electromagnetic radiation and "mechanical" interactions. Thus, in an elastic collision, for example, negatively charged electrons in a target react initially to their counterparts in the colliding object and visa versa. The resulting induced polarizations ultimately send the much more massive nuclei of the target chasing their bound electrons and running from the much more massive nuclei of the colliding object. The nuclei of the colliding object also chase their now rebounding bound electrons and run from the nuclei in the target. The net result is the transfer of energy and momentum from the colliding object to the target object. Energy and momentum are conserved in the process because energy and momentum are conserved in electromagnetic interactions and the collision is exclusively such an interaction.

14. Postscript, Application of the Concepts Above -- Rationalizing Surprising Phenomena in Astrophysics

a. Comments on Quasars, Active Galactic Nuclei and Cosmic Jets

The galactic structure suggested in an earlier section would simplify and rationalize the mechanism by which Quasars and Active Galactic Nuclei are powered and the process that leads to the formation of cosmic jets. The central region of any extended stable rotating disk-like structure containing the mass of a typical galaxy would be the focal point of all of the in-falling gas and other matter from above and below the disk. The individual molecules of the gas and the components of other in-falling matter would be accelerated to relativistic speeds and form jets moving toward the center of the distribution.

If the center of the distribution contains massive objects, then the accelerated jet (or jets if mass is located both above and below the plane of the galactic distribution) would strike these objects and the jets might very well collide with themselves and thereby create a small, intense source of radiation release and release of other energy very much like a collision zone in a particle accelerator (although with gravity providing the energy gradient that accelerates particles rather than electrical potential).

If the very center of such a distribution is largely empty (which one would ultimately expect to be the case in many systems given that the center of the system is exposed to an intense and unequal bombardment whenever the density of matter on either side of the plane of the system is different), then a result would be jets of material objects flowing inward and then outward at relativistic speeds through the center.[221] The outward bound jets would decelerate as they rose in the gravitational "field" of the galactic distribution, returning much of the gravitational potential energy they acquired in falling inward. Initially, the deceleration of a relativistic jet would have only a modest impact on its perceived "speed". Instead, the components of the jet would simply lose the augmented mass gained in approaching the speed of light. Ultimately, however, the deceleration would produce a material reduction in "speed" so that the jet would come to a transitional zone where particles would bunch and interact to create an energetic collision zone.

The mechanism suggested above would normally not produce symmetrical dual jets, but, instead, a variety of different possibilities exist depending on the arrangement of mass both above and below the plane of the galaxy. Dual jets would appear to be possible if, for example, there were mass above and to the left and below and to the right of the plane of a galaxy.

The mechanism above would also result in a focusing of electromagnetic radiation that is approaching from above or below the plane of the relevant system. Viewed along or very close to along the rotational axis of such a system with an empty core, the system would appear to have a central region of intense radiation generation, when, in fact, the

---

221 If one observed such a system from above or below the plane and along the axis of revolution of the system, the system would have much in common with the system created within a cathode ray tube. The author believes that such tubes generate duel jets (accelerated electrons in one direction and positive ions of ionized gas in the other) and that, viewed from the right perspective, one would experience extremely energetic phenomena even from objects at tremendous distances.

center of the system contains "nothing." The radiation that appears to have been generated from the system would actually have been generated elsewhere. Further, as one viewed such a system from along its rotational axis, it would appear to be generating immense amounts of radiation and could be observed at tremendous distances (just as quasars are). It would be a mistake, however, to assume that the flux of energy received at a particular point located along the extended rotational axis of such a system was representative of the radiation received at each point on a hypothetical sphere at the distance of this point from the relevant system. Further, if the source of the radiation "accelerated" by such a system was much closer to the system than the relevant observer, then the "blue" shift of radiation as it "fell" into the system would be less than the "red" shift of the radiation as it moved away from the system so that the "gravitational" red shift of such systems could be quite substantial. This red shift, if interpreted as indicative entirely of the distance to the object as is currently the case, would make these objects appear still more energetic and puzzling.

b. The Twin Paradox Revisited

Finally, let the author suggest a somewhat more realistic "twin" "paradox" experiment in an effort to highlight the problems inherent in ignoring the mass-equivalence of the potential energy that must be present in the fuels that are required for such an experiment. Begin with two rocket ship assemblies located adjacent to each other. Each has, as its core, a rocket that has a rest mass of 1,000 kilograms. With each core rocket are the following:

- Assembly 1: Sufficient fuel (that burns completely as all fuel considered in this problem) to accelerate the core rocket to a speed of .999 percent of the speed of light

- ◆ Assembly 2: A further subunit with additional fuel that would allow us to decelerate the core rocket plus acceleration fuel assembly from above back to rest; and
- ◆ Assembly 3: A further subunit containing the fuel that would allow us to accelerate the core rocket plus acceleration fuel (Assembly 1) plus the deceleration fuel assembly (Assembly 2) up to .999 percent of the speed of light.

The author calculates the mass of Assembly 1 -- the mass of each rocket plus the fuel to re-accelerate the rocket to .999 percent of the speed of light -- to be 22,366.27 kilograms using the formula:

$$m(v) = \frac{m_o}{(1-v^2/c^2)^{1/2}}$$

The author calculates the mass of Assembly 1 plus Assembly 2 -- the mass of each rocket plus the re-acceleration fuel plus the deceleration fuel -- to be 500,250.14 kilograms

The author calculates the mass of all three assemblies -- i.e. the mass of the rocket plus the re-acceleration fuel plus the deceleration fuel plus the fuel to originally accelerate the assemblies up to .999 percent of the speed of light -- to be 11,188,731.00 kilograms or 1.119×10^7 kilograms[222].

---

222 For a more rigorous analysis of the "explosive" growth of the mass of the fuel required to accomplish the sequence of events that set up the "twin paradox," the reader is referred to the discussion of the "Photon Rocket" at French, A.P., Special Relativity, W.W. Norton & Company, Inc. (1968), pages 183-184. As noted in that discussion the fraction "f" that a payload may be of the total mass of the best conceivable rocket -- a rocket with the optimum exhaust velocity of electromagnetic radiation -- is a function of the maximum speed of the rocket's payload. Significantly, for purposes of the present discussion, one applies the 4th power of that fraction to the initial mass of the rocket (including its fuel) to determine amount of the initial assembly that may return as the final

Of course, the author has made artificial assumptions in these calculations. Specifically, the author has assumed that none of the fuel assemblies interact gravitationally and also the author has allowed all of the energy in our fuel to be carried away by the particular assembly involved without considering the inevitable energy to be imparted to our exhaust. Nevertheless, this analysis highlights how rapidly universal mass grows as we contemplate motion within a universal system at high relative velocities. Given that the mass of the earth is 6.0×10^{24} kilograms, it is not hard to imagine how the total mass of the universe could get out of hand if we contemplate rapidly moving objects on the scales of galaxies (and especially if even a single galaxy contained a black hole at its center.

---

payload (assuming that the payload is brought to a stop when it returns to its origin, which, of course, significantly increases the amount of the required fuel beyond the amount required in the discussion above). The example in the referenced text discusses a time dilation factor of 10 (which equates to a speed of about .995 of the speed of light) and indicates that only 1/100,000 of the original assembly could return to the point of departure and be stopped there.

www.ingramcontent.com/pod-product-compliance
Lightning Source LLC
Chambersburg PA
CBHW051452170526
45166CB00001B/218